PRACTICAL
FOOD SMOKING

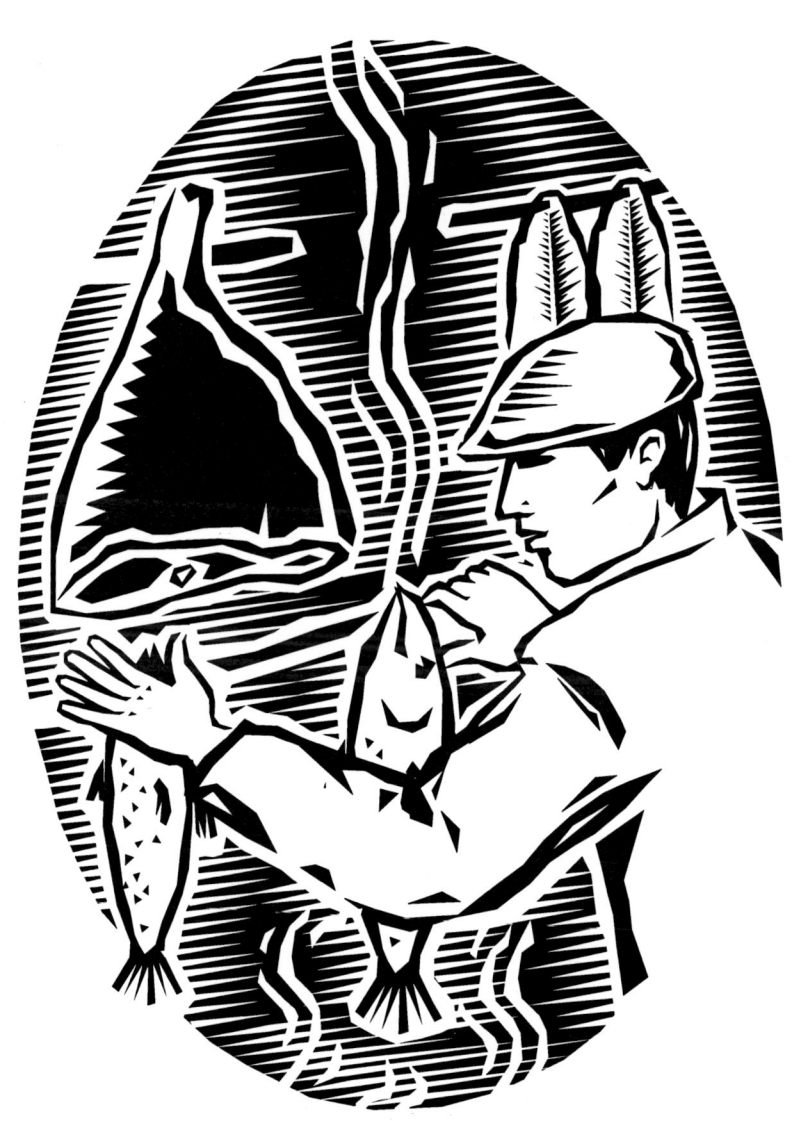

Practical
food smoking
~ A ~
Comprehensive Guide

Kate Walker

The Angels' Share is an imprint of
Neil Wilson Publishing Ltd
The Pentagon Centre
Washington Street
GLASGOW
G3 8AZ

Tel: 0141-221-1117
Fax: 0141-221-5363
www.nwp.co.uk
info@nwp.co.uk

The moral right of the author has been asserted.
A catalogue record for this book is available from the British
Library.
ISBN 1-897784-45-7

© Kate Walker, 2006
First published 1995
Reprinted December 2005 and February 2006

Illustrations by Doreen Shaw, Glasgow

Typeset in Plantin

Printed by Digisource

Contents

Introduction

URING RECENT years there has been a revival in the forgotten rural arts and a striving by some for self-sufficiency. Revolutionary changes have occurred in our society — more women go out to work, and with the increased pressures of coping with two jobs, the traditional skill of food preparation is no longer considered essentially a task for the female. The increase in the number of cars on the road have made it necessary to restrict parking outside high-street shops. Shopping is now being done less frequently and usually at one location, the supermarket, where ample parking is provided.

This 'rise' of the supermarket represents the levelling of food standards to a bland mediocrity where easily prepared dishes are available, and where quantity and continuity of supply are paramount. Unfortunately, it has also introduced food production practices which benefit the shareholder rather than the customer who has to buy what is available — even when he or she knows that the produce may not be as wholesome as is desired.

Problems in the poultry and meat industries have stimulated an increased interest in 'real food' and the thinking consumer is demanding access to facts about food production. The result of this is that a gradual change in eating habits is now emerging. Convenience foods are now found wanting and are being relegated to their rightful place of excellent emergency stopgaps. Accordingly, with interest in some of the more traditional food preservation crafts now rising, the process of food smoking, where an enhanced flavour is imparted to food by exposure to low temperature smoke in a natural manner, is becoming more popular at home and in business.

The fact that food cannot be smoked if water retention chemicals have been included in the animal feed is a safeguard against harmful substances being present and building up in the animal's flesh texture. The reason smoking shows up the presence of harmful additives is because salt draws out the moisture released by the process of osmosis, and this can be reversed to increase the water content and, consequently, the selling weight of the produce. There is a legal requirement to declare the water content if it has been added, but if this is brought about by doctoring the animal's feed, then no such declaration need be made. Some of the highest water contents are present in food imported from some continental countries.

The customer is becoming more aware of some of these facts, and greater demand for choice is bringing a further change in the structure of supermarkets where we now see counters springing up to cater for unwrapped produce. The rise of the American-style shopping mall, within which many smaller specialist shops are to be found, is also becoming a feature of today's urban environment. This represents an encouraging swing away from mindless buying, and provides a challenge to the producer with an eye on quality. Profit is a necessary part of any business, but not to the extent that the customer is duped. It will hopefully mean we can expect fresher food if buyers find their suppliers locally and deliveries become more frequent rather than distributing from central collection points.

I have written this book because there are many who would like to smoke food, but do not know how to go about it. The subject is complex and absorbing, and always full of new facts as more experimental work is done with a wider range of foods. It is unfortunate that little distinction is given to the labelling of smoked food to assist the public in the recognition of the difference between dyed, or artificially smoke-flavoured produce, and food properly smoked over a period of time where a combination of salt and cold smoke produces a better flavour and extended shelf-life.

This question of shelf-life is interesting because the image of smoked food being long-lasting and not requiring refrigeration is a leftover from the days when smoking was done for preservation. The method used is, in fact, the same for both preserving foodstuffs and giving enhanced flavour, but it is only the latter in which we have an interest today.

The methods given in this book therefore apply for preservation as well as flavour, but purely for the purpose of preservation a longer time in brine to increase the salt percentage is needed, as well as longer time in the smoker to reduce the overall moisture content. For an in-between requirement when the food is intended for reconstitution, salt levels are increased to allow greater transportation time, but not to the point that the foodstuff is completely dried out.

Anyone with an interest in the flavours of food should find something of value in this book which will serve not only the domestic household, but also the small business as well and to this end, I have looked at the commercial aspects of producing smoked food in order to give some guidelines for those who may wish to take their interest further.

Kate Walker

Chapter 1

A Brief History of Food Smoking

THE HISTORY of curing and smoking goes back some way and there is not a great deal known of how it all actually began. A lot of valuable information has been lost because early methods and recipes were handed down verbally, and as the need for people to preserve food has decreased, many facts have disappeared.

Food is necessary to survive, and man's instinct has always been to store enough food to last over the dark, cold days of winter. Primitive man, after bringing in meat to store for the winter (the 'autumn kill'), probably hung his meat simply to prevent animals devouring it, and then found that those joints exposed to the smoke of his fire remained in better condition. We know that once timber and stone houses were built, the rafters were used for hanging food because the smoke from the fire drifted upwards to the roof outlet. Possibly, too, coastal people might have become known as having better winter meat than those living inland because they used sea water rather than river water in which to wash their kill.

An interchange of provisions between these peoples may have resulted in a communal knowledge for meat and game to be kept in salt for quite long periods, then hung up in the roof space above the fireplace.

Nowadays, we are no longer in a position to maintain this type of store, but with rising costs and the improvement in freezing techniques, more families are buying foodstuffs in bulk, when prices are lowest, and then storing this in a deep-freeze for future use.

Some cuts can easily be cured and smoked before freezing but the difference today is that food is not being cured and smoked as a means of preseravtion, but rather as an enhancement to flavour. This fact in some way influences the process because our modern, sophisticated palates would reject the very salty and dried texture of ancient man's winter stores. Having said that, it must be understood that although smoking imparts a longer shelf-life, it is still necessary to treat smoked products as raw food, still requiring refrigeration or freezing for storage, unless the products have been smoked specifically for preservation.

The use of smoking for preservation is still of prime importance to under-developed countries where widespread refrigeration is not yet common. The raw material is often available, but it will not keep beyond a day or so, unless given extended life by preservation which can allow time for distribution and storage. Rural communities can benefit most from this type of processing because the lack of shops supplying protein is very real. Smoking can provide a means whereby small local communities with access to fishing or other raw materials can catch and smoke more than can be eaten in a day, and then sell any excess to inland communities, thus providing a means of strengthening their local economy.

It is of interest that smoking methods differ from country to country and on the European continent salting has not been a part of the process, thus giving a product which either has to be cooked until dry or packed in salt. The loss of essential proteins and fatty acids has resulted, and the product does not look attractive. The methods I go into in this book will bring about presentable and attractive smoked food which can be consumed at any time of year.

Chapter 2

What is Food Smoking?

SMOKING IS a form of preservation which reduces the moisture content of food and gives a certain amount of protection against bacteria, due to the chemical changes that take place within the flesh as a result of brining and the effect of wood smoke on the salted flesh.

The terms 'cold' and 'hot' smoking have led to a certain amount of confusion. Cold-smoking is the true smoking and the method by which food is changed in both colour and flavour. It is carried out in a temperature range, usually between 21-31°C°/70-88°F, but ideally at about 25°C/77°F. If higher temperatures are used, various components within the wood, sawdust or wood chips are vapourised and a hard skin forms on the food surface which resists the wood smoke penetrating the food, and an inferior final product results. In Europe the more usual method of smoking is to apply some heat to dry the fish or meat initially, and then cold-smoke it. This gives a dark-coloured product with very little smoke flavour apart from the surface area near the skin.

Similarly, many operators do not salt before smoking and therefore, the chemical changes which give enhanced flavour and a characteristic golden colour, do not take place. The effect of hot-smoking is that the level of carbon particles or traces of aromatic hydrocarbons, after the process is complete, will tend to be higher than cold-smoked products because low temperature smouldering of wood will release vapour rather than particulates. Having been cold-smoked some foods are then hot-smoked to enable them to be eaten without further preparation. This final process is usually a short one in order to kill bacteria or to make the flesh tender enough to eat. If no cold-smoking time has been given to food first, it would be more accurate to describe the food as barbecued and it is one of the reasons why smoke ovens are more in demand because they serve as a means of preparation for barbecues to ensure the food does not take a long time to prepare, but is still edible.

The lack of understanding of the smoking process has caused confusion in the minds of most laymen who consider smoked food solely as as preserved food, and therefore, requiring little or no care. As has been explained elsewhere, generally we smoke food today for enhanced flavour, and it is therefore important that this side of the process is given priority and that the time taken for the whole process is biased towards this factor. Time, however, is money, and many processors use high-velocity air flow to give the simulated texture of a smoked product without the chemical changes taking place, relying on the salt content to give both flavour and sufficient shelf-life to sell on the product.

If air is driven through the smoker above a certain rate, the smoke vapour will not settle on the surface of the food and no flavour develops. This is also true of the change in colour, but this can be altered by the use of dye whereas the flavour will not be present unless introduced artificially. Traditional processors throughout the world are aware of the threat to their livlihood from the competition of some modern methods of production. At present the legislation to differentiate between the methods is not sufficient.

I have explained that all products are cold-smoked, but that it is not necessary to hot-smoke all food. Cold-smoked products can be offered to the consumer to be cooked by normal methods without a loss of final flavour, and as long as the customer has instructions for cooking, there is no risk involved. But, due to the fact that some processors have improperly offered cold-smoked, large rainbow trout for consumption without cooking, there is good reason in the mind of the consumer to be aware of such practices.

These processors have used a raw material of much lower cost, in order to produce something similar to smoked salmon which yields a much higher profit margin. The texture and appearance are acceptable, but this practice is dangerous, due to the possibility of botulism. Warnings have been given but have had to be carefully handled because the media might take the matter up and inadvertantly produce an effect just as harmful to the fish-farming industry as an actual incident arising from the public eating infected fish product.

Fish from fresh water must be cooked in some way, and treated with acid or hot-smoked. The reason for this is that all fresh water has contact with the ground in some way. The ground contains spores of the bacillus Clostridium botulinum and these can be transferred to fresh water, and so into fish. If food in which the spores are present is improperly handled, a deadly toxin forms. Heat kills this toxin, so it is essential to cook food.

On the other hand, some species of sea fish such as the Atlantic salmon can be safely eaten after cold-smoking because their life cycle is different; they enter fresh water for a limited period only, or they are caught in sea areas which are relatively free of foreign bodies. Another problem area is that the United States prevents the sale, when cold-smoked, of the Oncorhynchus species (sockeye, coho, chinook, chum, humpback) caught in the Pacific where Clostridium botulinum is present. However, it is still possible to import these species to Europe and sell them as cold-smoked produce.

When salted food is hung in a stream of cold smoke, the action of enzymes causes changes to take place in the food giving a characteristic golden colour and enhanced flavour due to a slight loss of natural moisture. Salt is introduced into flesh as dry salt or as a brine solution by osmosis. This is a natural process which states that when a substance, at a cellular level, is put into a saline solution or covered in dry salt, then the intracellular salinity will attempt to equalise with the saline concentration outside the cells. Thus, the process of osmosis will attempt to reduce the surrounding salinity by releasing some water and taking in some of the salt. The net result is an increase in intracellular salinity, and a slight decrease in extracellular salinity. In effect, the cellular structure is diminished in the process.

When hardwood sawdust or chippings smoulder at a low temperature below 26°C/80°F, acetates and aldehydes are released in a

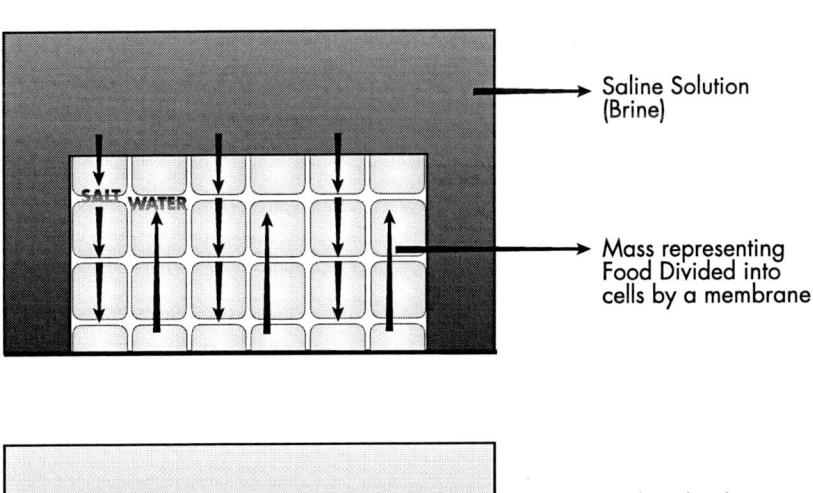

Saline Solution (Brine)

Mass representing Food Divided into cells by a membrane

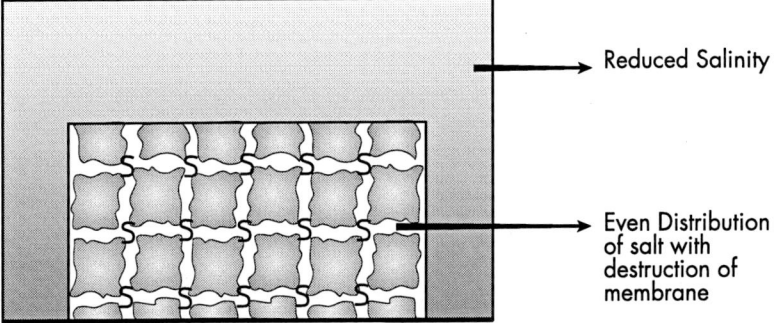

Reduced Salinity

Even Distribution of salt with destruction of membrane

pale vapour which settles on the salted food, combining with the salt and, through the action of enzymes, these disperse throughout the food altering both its colour and flavour.

These last two paragraphs oversimplify a process that takes time and involves the complex interchange of enzymes and micro-organisms. As already explained, it cannot be hurried and today, when 'time means money', the buying public have a right to demand that quality for the extra money they must pay for smoked produce.

In order to successfully smoke food, it is not necessary to understand the intricacies of the process, but from a health and safety angle some basic facts should be known. All organic material such as meat, game, fowl and fish have enzymes and micro-organisms present in their make-up. These are essential components in the biochemical and physiological processes which occur in living tissue. When slaughter takes place it is impossible to stop the resultant activity of these substances and the flesh continues to metabolise oxygen. In fact, we want these reactions to continue to a certain degree. The flesh of anything newly-killed is tough and tasteless, but held in cool, but not freezing temperatures, the enzymes will begin to help soften the flesh, and consequently make it more palatable — in other words the flesh will mature.

The flesh of land animals takes longer to reach this matured state than fish. The actions of enzymes and oxygen in meat fat are quicker and tend to turn it rancid within a short period so that surface fat often has to be removed. Modern abattoirs deal with this critical ripening period under very strict regulations. The maturing time for beef should be longer than for pork, mutton or lamb. Any flesh not allowed this maturing time will not smoke, because osmosis does not take place in immature flesh. As a guide to maturing times, a minimum of one week should be allowed for beef, venison, pork, mutton and goat. Fish and shellfish, on the other hand, should be used as soon as rigor mortis is over. If these conditions are maintained in the raw material then both salt and smoke can penetrate. For the same reason of immaturity the flesh of calves, very young lambs and kid is not smoked because there is neither texture nor flavour, and no salt can be taken in.

Once started, the breakdown of tissue by enzymatic action must be stopped at some point to avoid spoilage, and this is now done by controlling the temperature, by either freezing or cooking.

The other group of natural substances that must be considered are the micro-organisms which can be divided into three categories, (these are dealt with more fully in chapter 10).

1. Moulds
2. Yeasts
3. Bacteria

Many interesting books, pamphlets and papers are available on these three complex subjects, but for our purposes a bare outline only is necessary.

1. A mould is a fungal growth which is visible and easily recognisable. It is a surface growth and is usually found in damp conditions. Boiling destroys moulds and their growth is arrested by freezing below 0°C/32°F. (When I later discuss the storing of smoked food, particularly ham, dry cool places are recommended to avoid the growth of mould.)

2. Yeasts are living plant organisms which, with the help of enzymes, consume natural starches and sugars to produce carbon dioxide and alcohol. This process is called fermentation and is not always as visible with regards to food as it is when beer is being brewed. It can therefore spoil food without notice. A temperature of -18°C/0°F renders yeast inactive as will 121°C/250°F. In other words, deep-freezing or cooking.

3. The third group, and the one with which most people are familiar, are bacteria. Not all bacteria are malignant, but in the context of the processes of curing and smoking, those which cause the decay of flesh must be destroyed. Bacteria will remain dormant below 0°C/32°F and by exposing them to a temperature of 90°C/194° most bacteria will be destroyed, but in order to ensure total destruction, they must be subjected to a temperature of 121°C/250°F.

The following details of the smoking technique may appear to be a contradiction of all I have written above because of the requirement to keep temperatures low during smoking, but the important factor which explains this apparent anomaly is the use of salt.

Chapter 3

Selection of Raw Materials

RAW MATERIALS for smoking must be of the best quality because you will only get out of the smoker what you are prepared to put in. In order to prepare these materials correctly, they must be correctly salted, otherwise there will be no interaction with the smoke. Materials in a condition which prevents correct salting by the process of osmosis, will inhibit the smoking process. Remember, without an even distribution of salt in the tissue, there can be no interaction with the smoke.

There are some conditions under which osmosis will not take place:

1. The meat has not been through the condition of rigor mortis.

2. The flesh is too immature, i.e. from animals which are too young.

3. Undue stress has occurred at death with a resultant high adrenalin content.

4. The atmosphere is too cold, making the brine chilly.

5. The presence of polyphosphate in the flesh, sometimes included in the feed.

6. Other substances such as bicarbonate of soda have been present in feeds to retain water and create extra weight in the flesh.

There are three main divisions of raw material — meat (such as mutton, lamb, pork, goat and venison), fish and poultry. All these materials must come from animals which have fed properly and are thus in good condition when killed, then allowed to go through rigor and have the maturation process take place. Bearing in mind the need for osmosis during processing, feeding should be as nearly natural as possible avoiding the introduction in feed pellets of substances not normally part of the diet of that species, e.g. animal protein for a herbivore or fish meal for a vegetarian. This is not for ethical reasons, but because the defence mechanisms inbuilt in a species are not designed for combatting bacteria normally associated with another species. Admittedly, it can be difficult to find out from what materials a proprietary brand of feed is actually made.

Today's intensive production means feeding an animal a concentrated diet which will enable a specified weight to be gained in the shortest time. This is often carried out with a minimum of activity so that although there is bulk created, there is little muscle tone. Unless a starvation period before killing is allowed, the tone will be even worse and the flesh will be softer and less tasty. Land animals are usually allowed a longer time to reach the required killing weight than poultry and farmed fish, and have more access to activity, but they still require pre-killing starvation. If an allocated starvation period before killing and a maturing time after rigor has been completed, this will allow good muscle tone to develop. Pork presents a good example of how husbandry can affect the end-product. Pigs are temperamental and if kept happy while alive, they will produce good meat. They do require exercise, somewhere to root around and, if possible, somewhere to wash. The results always outweigh the extra effort. Careful handling when being taken to market is essential. If a pig is in distress, too much adrenalin is released into the system and the flesh after slaughter will never be tender enough to eat.

Meat

When the time to kill arrives, we have to differentiate between the methods used for meat, fish and poultry. Abattoirs have come a long way and now despatch beasts with a minimum of stress, which could be further reduced if waiting time could be abolished with animals arriving just prior to slaughter. For those arranging the slaughter of beasts, certain conditions must apply by law after slaughter, but it will be up to you to ensure that rigor takes place naturally and that a maturing time of around eight days is allowed before brining.

The decrease in pH (the degree of acidity/alkalinity) due to an increase in acidity which follows rigor allows chemical and enzymatic changes to take place, and those, in turn, are what creates the increase in flavour and development of texture. A biochemistry lesson is not required to prove that this maturing period is necessary. Present-day meat on sale in supermarkets is often immature, and it does not come as a surprise that many people are shunning meat flesh because they have never experienced its true flavour.

Fish

There exists a general callousness to the treatment of fish, perhaps because they live in their own unique biosphere. It is more difficult to regulate the conditions of their capture and killing due to the remoteness of the point of capture and the subsequent trans-portation time while in ice. Fish brought from the deep water too quickly suffer from what humans would call 'the bends' and show signs of haemorrhage around the eyes where the flesh can be suf-fused with blood. If then being thawed from frozen, the flesh will go off very quickly and be very soft. Suffocation of fish at the bottom of the net by the weight of overlying fish is another problem because the stress which this induces makes the flesh very soft, and it is thus unable to take in salt. Many fishermen however, land perfectly good catches, and if you are in a local port, buy from the smaller boats as the above problems will not be so apparent.

Today, a high percentage of fish used for processing is farm-reared. Some farms are excellent at producing fish resembling the wild equivalent. Others produce soft and flabby fish due to the wrong ratio of feed to movement, too short a starvation time before harvesting, and the use of a method of killing which creates too much stress. At present, admittedly, there is no ideal way to despatch large numbers of fish and hitting each one on the head is out of the question, although this remains the most humane way.

Some methods are worse than others and the use of carbon diox-ide introduced into the killing pond or container creates the most stress as the fish go berserk in their search for uncontaminated, oxy-genated water. They adrenalinise themselves to provide the natural speed to escape, and this ruins the flesh.

A widely used method of killing fish is to allow them to bleed to death by cutting the gills or slitting the main artery at the tail. Without first administering a stunning blow, many people consider this form of bleeding cruel. Considering that the fish have to be han-dled prior to bleeding, there is really no excuse not to stun the fish first. The advantages of bleeding are enormous and this should always be undertaken because, apart from doing away with a possi-ble source of infection from bacteria, the flesh is kept free of bruises or blood marks after smoking.

Fish usually go into rigor more quickly than land animals, and farmed fish will take much longer than wild, but in either case, rigor must be complete before starting to brine.

Poultry

The final group is poultry and never has the mass production of a food introduced so many doubtful factors. This is not the platform to air my views however, and only those aspects which will restrict proper smoking are of interest here. The feeding and feedstuffs have already been mentioned, and must be considered. The normal killing of poultry today is done on a line where their heads are chopped off. The birds are then put through a scalding tank, plucked, eviscerated and put into the freezer as soon as possible. Apart from the inducement of stress, putting birds into hot water to ease feather removal will inhibit salt going into the flesh. Being frozen so quickly stops the natural rise in acidity so the salmonella bacteria have nothing to inhibit their growth. As rigor is not allowed to set in, there is no continuing fall in pH, so no flavour or texture develops. The result is a cheap product with very little taste. This tastelessness is one reason why smoking poultry can be so useful, and all the mentioned pitfalls can be avoided if birds are given some movement during their lifespan, killed humanely, and allowed to go through rigor before processing.

When a creature dies or is killed, the meat is tough and tasteless. In life, biochemical processes are constantly changing, producing energy from the consumption of food. On death, the blood stops being circulated, so there is no more oxygen available or energy produced and at this point, the muscle will die and rigor mortis follows. The process is more complicated than that, but for the purpose of food smoking it is adequate to merely recognise that it happens. Rigor mortis is recognisable because the creature becomes rigid, and handling it while in a state of rigor can damage the flesh. The onset of rigor can be delayed by the application of ice or a freezing temperature. It will also be slower to commence in animals with good glycogen reserves or, put more simply, animals which have enjoyed good health and eating.

Rigor in fish is similar to land animals, but may occur at higher temperatures due to the catching conditions. In extreme cases, when the temperature is in the region of 17°C/62°F, the contractions can be so strong that the connecting tissue, which envelops the muscle layers, is torn and the muscle tissue flakes separate, leaving gaps in between. This is known as 'gaping' and presents a problem if the fish is to be sliced when preparing smoked salmon or other fish. If too little starvation time has been given to farmed trout, for example, or too much stress has been induced at death, instead of gaping, a sticking together of the tissue flakes occurs which renders the flesh unfit to slice.

Fish, although not traditionally 'hung', are better on completion of rigor. In wild salmon a creamy substance called the curd is exuded and is reabsorbed during the 24 hours after rigor is complete with a consequent increase in flavour.

Game birds are hung for considerable periods which vary from two to 10 days, but when smoking, it is better not to extend this period beyond two days, because the flavour which manifests itself after smoking is considered by many to be too strong.

Any other raw material likely to be handled will fit into one of the above categories, and can be judged as fit, or not, to be used in the smoking process.

Chapter 4

Brining

SALT FORMS one half of the smoking process, the actual smoke being the other. It is therefore important that the salt is of the best quality available. Salt is one of the most valuable substances known to man and, historically, it can be linked to the growth of civilisations because the presence of this commodity has meant wealth. It has been sold, bartered and its use taxed: all three represent economic advantages where deposits have been found. We tend to treat salt with little respect, and if smoking is being contemplated as a business, it would be advisable to change that outlook.

Salt always originates from the sea, but can also be obtained from deposits left behind after the sea has retreated during geological upheavals. The three main kinds of salt can be categorised as:

1. Sea salt
2. Rock salt
3. Vacuum salt

Sea salt is made by filling ponds (or 'pans') with sea water which evaporates due to the effect of sun and wind. A suitable climate for this is needed, and restricts the areas in the world in which good sea salt can be found. The purity of the sea water being trapped prior to evaporation and grading, is also important. Suitable sea water is becoming more and more difficult to find, due largely to pollution.

Rock salt is obtained by mining the beds of fossil sea salt accumulated many thousands of years ago when large sea areas evaporated. These deposits tend to be free from modern-day pollution.

Vacuum salt is obtained by either dissolving rock salt in situ and pumping it as a solution to the surface, or by mixing rock salt with water after the mining process. In either case, once on the surface it is thermally processed in large evaporators during which the salt is purified and dried. It can then be re-crystallised or left as a fine power, which is the form used in most domestic applications. This salt is kept free-running by additives, usually introduced at the solution stage of processing.

Rock salt is the most economic to use, and can be bought in different crystal sizes to suit the product and process. The minute amounts of trace elements found in natural rock salt are also a bonus and add an imperceptible, although valuable, addition to the smoking process and this, in turn, is of nutritional value to the consumer of the smoked food.

When selecting the type of salt read, the chemical breakdown on the manufacturer's label, and select the one with the *least* amount of magnesium sulphate and calcium chloride, because even if the amounts are very small, they tend to attract moisture from the atmosphere, and this can encourage spoilage by the development of parasitic and fungal growths.

The temperature of cold smoking is so low that the oil of a particular type of wood which might impart flavour, is not volatilised and does not influence the flavour of the end-product. It is often necessary to add a little something to make the end-product more characteristic, but the labelling regulations make it difficult to have a 'secret' ingredient! These additions are added into the brine and take the form of herbs, spices, wine, beer, alcohol or anything which you feel will give a pleasant taste, providing it does not inhibit the actual smoking process. It is always good to take careful notes when experimenting to reproduce the selected end-product, and then to make extensive shelf-life tests to ensure that further changes do not occur after a period of time.

For instance, the addition of some peat to the sawdust should be carefully undertaken because if it has been formed from soft wood it will give a very acrid final flavour and tends to burn at too high a temperature, giving off some of the less acceptable components produced during combustion. The harder the wood, the longer and cooler the burning will be, and it is sometimes necessary to mix the sawdust of various oak and beech woods to obtain a stronger colour to the finished product.

When the term *brining* is mentioned, the use of salt, either dry or in solution, with or without added ingredients, is what is meant. A salt solution with sugar, spice or herbs, and saltpetre is more often called a cure or pickle. The choice of method is dependent upon:

1. The raw material.
2. The length of time it will remain in brine.
3. The length of time it is required to be kept before eating.
4. Alternative methods of preservation, i.e., refrigeration.

The brining solution and time is only strong enough and long enough to bring about the desired end-product which will then be eaten or frozen. If preservation is the sole aim, the curing process must be longer and more salty. The modern palate would reject the salty, dried-up food that gladdened the heart of primitive man in mid-winter.

Sugar is added to brine because salt tends to harden flesh whereas sugar tenderises and fixes colour. It also provides a medium for the bacteria necessary to break down the sugar into organic acids, some of which give pleasant flavours.

Saltpetre is added to brines to inhibit the growth of bacteria and to help keep the natural pinkish colour of flesh. Do not use more than is stated in the recipes because its chemical action is desired only to a small degree. The degree of saltiness of the brine can be a personal decision, but it must be strong enough to do the essential work of curing the food. By osmosis, brining commences the process of weight loss as water from the flesh is drawn out.

Dry-salting tends to give greater weight loss than brining, and with fish this can be as high as 9%. This is counteracted by less weight loss during the smoking. Total weight loss from cleaned and gutted fish to the end of the process should be about 17%. For the beginner, it is advisable to weigh the material after each step, but as more experience is gained, the look of the finished food will suffice. The weight loss with fowl and game birds works out at about 20%, and with meat and venison, nearer 25%.

There are two schools of thought with regards to brining. Some people prefer strong brine and a short brining time; others prefer a weaker brine with a longer brining time. Although no hard or fast rule prevails, the stronger the brine used, the less time is required. At 80% salt solution (4.54 litres/1 gallon of water to 1.22kg/2lbs 11 oz of salt) efficient brining takes place with short brining times and using this as a starting point, personal experience and preference will result in the use of varying salt percentages.

The decision as to whether to use a brine or the dry-salt method is again a matter of choice, but if the ambient atmospheric temperature is high (over 26°C/80°F) the dry salt method will give quicker protection against spoilage. With refrigeration or cold room facilities however, this does not influence the choice to any degree.

With dry-salting some means of draining off the moisture extracted is required — usually a tray or slab with a drain. A layer of salt is spread on this and the food to be salted laid flat on top of this. Salt is then sprinkled generously on the upper surface of the food, and further layers of food and salt added until the batch is completed, ending with a layer of salt covered with a weighted wooden board. In the case of fish, sprinkle less towards the tail as the flesh is thinner and would become over-salted. When the food being dry-salted is not flat, it is not so easy to layer, but the individual pieces can be put into a container and salt sprinkled round about. Because of this difficulty in ensuring even distribution, the brine method has distinct advantages.

Making brine

To make a brine, use water that is as pure as possible, to which the required amount of salt is added and stirred around until dissolved. If herbs or spices are being used, boil them with a small amount of the water for 15 minutes or so, strain the liquid into the brine water and retain the whole spice or herbs by tying them in a muslin pouch which can then be put into the solution. (Depending on the final use to which the food will be put, the decision to add flavour in the form of herbs or spices, flavoured salts or wine will have been taken. All have a place in smoking, the end justifying the means). Chill this to approximately 3°C/38°F and put the food into this. Maintain at 3°C/38°F for duration of the cure. This temperature is important because it inhibits the action of bacteria and gives the salt time to act. If maintained at a higher temperature, spoilage will occur and at lower temperatures no osmosis will take place.

Brine solutions can be used more than once provided they remain in good condition. Brine for fish should not be used for any other type of food. A brine should not be used again if it smells sour or tainted, has blood in it from previous use, or appears ropey. Salt is relatively cheap and it is better to mix a new brine than suffer failure by using a doubtful brine again. If in doubt, don't use it!

When brining or curing over a long period, for example with hams etc, the solution should be overhauled on a weekly basis. This ensures that the cure is going ahead without a problem. When the container is opened it will smell pleasant with no hint of sourness or taint. The food should be turned and the brine stirred up to ensure even curing. If the brine shows any sign of being off, throw it away and make up another batch. It will be possible to tell if the food has developed any spoilage, particularly if unboned, because it will smell rotten and the appearance will be slimy or even greenish in colour. In this situation little can be done, but sometimes by removing the bones and any surrounding flesh, some part may be saved. Again, if in any doubt, don't save it.

When food is being brined it tends to float and some means of keeping the articles covered has to be taken. A well-scrubbed plastic board with a weight on top serves very well.

The brining process completed, it is important to allow drying time before the food is put into the smoker. This allows excess moisture to drip off and the surface texture to form. This salt gloss which gives the finished article a uniform appearance is called the pellicle and varies in thickness, depending on the conditions prevailing when drip-drying and when in the smoker. If it is thick, it is usually removed before selling on.

Brine salinity can be measured directly by the use of a brineometer (or salineometer) which gives the percentage of salt in the solution when the instrument is allowed to float in the brine. When a brineometer is unavailable, the following table should be consulted to give the correct concentrations.

Brine strengths at 15°C/60°F. Weight of salt required for relevant percentage concentration

% solution	Imperial gallon	US gallon	Litre
40	1lb 3oz	1lb ¾oz	118g
50	1lb 8½oz	1lb 4¼oz	150g
60	1lb 14oz	1lb 9oz	185g
65	2lbs 1oz	1lb 11½oz	206g
70	2lbs 4¼oz	1lb 14oz	225g
80	2lbs 10¾oz	2lbs 3½oz	268g

The brine pump

This is perhaps the best place to make mention of this item. The use of a brine pump, or more recently, the needle injection method of introducing salt into a joint is a helpful addition, but no substitute for the longer method of brining as described above. Water retention is the most serious factor in bad smoking, because it means the salt distribution throughout the flesh is uneven, and consequently, the smoke cannot combine with the salt properly and no effective smoking takes place.

There are times, however, when a brine pump is very useful, and in fact, desirable such as when smoking large turkeys, or when preparing joints with the bone in. In the case of a turkey a small amount of 60% brine solution is introduced to the thickest part of the thigh and breast. Similarly, if curing a large piece of flesh on the bone, taint is prevented by the contact with salt as the brine solution is slowly penetrating from the outside. By also introducing salt internally, the brine is working on the inside as well as the outside and this allows a reduction in the brining time. It ensures salt penetration into even the thickest parts of the flesh. Osmosis will also ensure an even salinity and a good flesh texture. Note that it is usual to scrape out as much marrow as possible from the bone, because it goes rancid much quicker than the flesh, due to its higher fat content.

A brine pump comprises of a number of parts which in principal are the barrel in which the salt solution is collected, and the needle, through which the solution is injected into the flesh. Once filled with brine, the needle is inserted into the part of the flesh which requires the salt and the plunger is slowly pressed in, while the needle is withdrawn. This ensures that an even distribution of solution is injected.

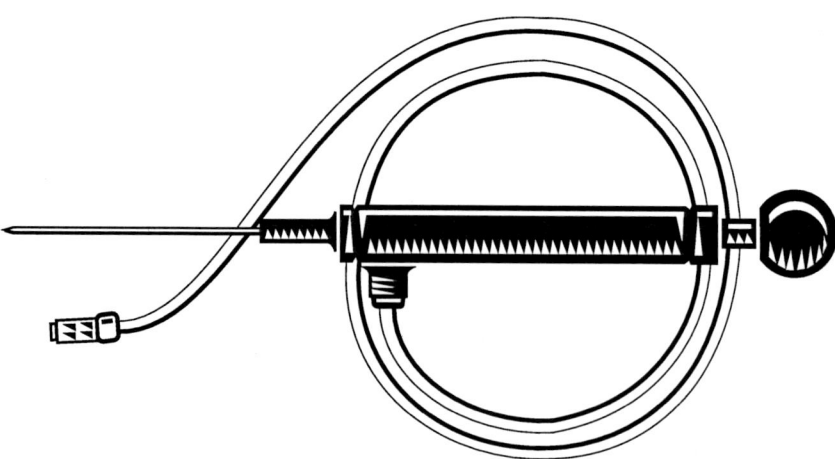

When the needle is extracted, press the hole to seal the flesh; this prevents the brine from leaking out. The process can be repeated in other areas of the flesh. Once completed, the bird or joint is then placed in brine in the usual fashion, but the length of time is cut by about half.

When using a brine pump ensure that the pump is sterilised before use, especially the needle which should be put in boiling water for at least one minute if no sterilising equipment is available. The same procedure is recommended after use.

Sample brines and cures

Note: where the quantities of water are not enough to submerge the food being cured, multiply the make-up in proportion until a sufficient volume is made.

1. Brine for hams, tongue, venison or other cuts of beef

> 20 litres/4½ gallons of water
> 1.1kg/2 lbs 8 oz salt
> 500g/1 lb sugar, — white or soft brown
> 15g/½ oz saltpetre
> 60g/2 oz pickling spice

Boil one pint of water and add the pickling spice. Allow to infuse for 15 minutes. Mix the salt and water. Add the drained pickling spice water, retaining the spices which are wrapped in muslin and then added to brine solution. When the salt is thoroughly dissolved, add the sugar and saltpetre.

Chill and then add the cut to be cured.

2. Brine for tongue, rabbit, pork fillet and small cuts of venison.

> 1.1 litres/5 pints water
> 180g/6 oz salt
> 45g/1½ oz white sugar
> One-eighth tsp saltpetre
> 4-5 bay leaves
> 1 tsp cloves
> 1 tsp allspice

Method as above.

3. Brine for chickens or turkey

> 4.5litres/1 gallon water
> 2 tbsp onion salt
> 1 large clove garlic
> 500g/1 lb salt
> 2-3 fl oz mild white vinegar
> 500g/1 lb light, soft brown sugar
> 2 bay leaves
> 2-3 cloves

Cream the onion salt with a little water. Peel and split the garlic. Mix the salt into the water. Add the creamed onion salt, vinegar, sugar and spices. Chill.

4. Brine for game and paté ingredients or sausages, liver, kidneys etc.

> 4.5litres/1 gallon water
> 500g/1 lb black treacle
> 2 tbsp onion salt
> 1 clove garlic
> 500g/1 lb salt
> 60-90ml/2-3 fl oz mild white vinegar or wine vinegar
> 6 juniper berries
> 2-3 bay leaves
> 2-3 cloves
> 500ml/1 pint red or white wine — or plain vinegar used
> with a dash of brandy to taste

Heat the treacle gently in its tin to make measuring and pouring easier. Cream the onion salt and peel and split the garlic. Simmer all the ingredients together. Stir well until the salt is dissolved. Chill.

5. Brine for salmon

2.25litres/4 pints water
500g/1 lb salt
1 tbsp soft brown sugar or demerara
2 bay leaves
5-6 juniper berries
½ tsp saltpetre
Rum (about 2 tbsp)
Mix as previously instructed.

6. Dry cure for spiced beef (enough for a 6-lb joint)

500g/1 lb coarse salt
1 heaped tsp saltpetre
2 heaped tbsp brown sugar
1 large tsp allspice
1 clove garlic (optional)
1 level tsp peppercorns
6 blades mace
1 tsp cloves
3 bay leaves

Rub all the dry ingredients together, then pound in the bay leaves and garlic (if used). Stand the meat in a large earthenware, plastic or glass dish and rub the spicing mixture thoroughly all over it. This should be repeated every day for a week taking the spicing mixture from the bottom of the dish and turning the meat twice.

Chapter 5

Smoke ovens and other equipment

SMOKERS COME in a variety of sizes and designs and your interest will centre around the particular smoking programme you want to follow, be it home-based or commercial. With all smokers the end-product is all that matters, and therefore it is essential to choose the smoker which will fulfil your requirements best. Do not take a sledgehammer to crack a nut!

What Do Smokers Do?

Smokers have to provide an adequate supply of smoke and air in the correct proportion at the correct temperature in order to carry out their function of cold-moking, hot-smoking and drying. To do this they have to have the following parts:

1. An *oven* of sufficient capacity to contain the foodstuffs suitably suspended or supported.

2. A *smoke generator* capable of producing a sufficient quantity of smoke for the maximum load of foodstuff. Also, a system which ensures that the smoke and air are evenly distributed around the food.

3. A *heat source* to produce the correct degree of heat when hot-smoking.

4. *Fuel containers* which hold the sawdust from which the smoke is generated.

4. A *temperature control device* to monitor and maintain the correct temperature of the burning sawdust.

What makes a good smoker?

A good smoker should be efficient, that is, it should use a minimum of power to create the necessary heat, air flow and smoke. It should be convenient to control and it should be safe in operation.

A Global II 250kg nominal capacity smoke oven. An oven such as this will cost in the region of £20,000 to £30,000. PICTURE: BILL MILLER

Modern technology provides the means of automatic operation, close temperature control, adequate protection, and allows for a sequence of operations to be programmed to take place without attendance. These are all good in their own way, but it should never be forgotten that the person in charge must know what the end-product should be and be able to recognise variations in the raw material. For example, salmon or other fish high in oil content require longer time to brine and smoke. Models capable of handling loads of a few kilograms are available for domestic and small commercial use while custom-built commercial models can deal with loads of hundreds of kilograms. There are obvious advantages in having as compact a machine as possible.

Oven fittings

The provision for hanging or supporting the food is important so that it is evenly exposed to the smoke. A variety of trays, bars, rods and hooks can be used, depending on the pieces of food involved, and these fittings are usually supplied by the manufacturer to suit the customer. The usual way of supporting these fittings is from a series of brackets fitted into the walls of the smoke oven. For larger machines these fittings can be used with a trolley which is manhandled into and out of the oven. To suit modern-day legislation the trolley goes in one door and out of another on the opposite side.

The oven walls are normally of hollow construction and the internal surfaces are of stainless steel for all commercial ovens, and/or galvanised steel for smaller smokers. The outer walls in large smokers are usually stainless steel, and in smaller ones, either stainless or galvanised steel or aluminium. The oven is fitted with a close-fitting rigid door or two doors with latching handles and these are effectively sealed by means of rubber strips.

The smoke generator

This, not surprisingly, produces the smoke for the smoking process. Apart from the sawdust fuel, a controlled supply of air is required for the slow combustion process required and this flow of this air is used wholly or partly to convey the smoke into the oven. The sawdust or wood chip mixtures are contained in a metal fuel box (or series of boxes in larger ovens), which have holes in their front facia to allow the entry air. These boxes are situated in a separate sealed compartment either below or to one side of the oven. A separate, sealed door is provided for this compartment and the floor or walls have louvred or screened openings in them to allow air entry. Depending on its size and arrangement a smoke chamber may have a heat baffle, which is a perforated metal screen to arrest sparks or debris, and a temperature sensor. A drip tray for the collection of tar and other debris might also be fitted.

The Gourmin domestic cold-smoker of 5kg nominal capacity, costing £189. This is ideal as a training oven to develop product and techniques. This operates on the natural draught technique.

How do I start up a food smoker?

In conventional smokers the sawdust is ignited by using a small blowtorch and playing the flame through the holes in the front of the fuel box until the sawdust is alight. By allowing about a minute per hole, the sawdust will glow red before settling down to a black smoulder which is the desired state.

Heat sources

To provide an input of heat to the oven for hot-smoking, drying, cold-smoking in very cold damp weather, or for oven start-up, an arrangement of stainless steel-clad electric resistance heaters is required. These are situated in the air stream to the oven, either below it or within its hollow walls. They are so arranged that the air stream is heated evenly. The kilowatt electric output is sufficient to maintain the required maximum temperature, but the rise in temperature is slow to ensure even heating of the food charge and to assist in achieving accurate control of temperature. Most smokers are provided with facilities for manually selecting the amount of heat required: low for drying and start-up, and high for hot-smoking.

Temperature regulators are usually fitted to select and maintain the required oven temperature to a close degree.

Air flow and smoke distribution

How does the smoke get into the oven? Unlike the haphazard methods of the past, in a modern smoker the correct quantity of smoke is conveyed from the smoke generator into the oven by means of a controlled flow of air. This airflow can be brought about in several ways:

1. Natural draught. In a domestic solid fuel fire a draught of air is produced by the burning of the fuel allied to the height of the chimney. This flows through the fire grate and hearth and up the chimney. In the same way the burning sawdust in a smoker can provide sufficient heat to induce a natural draught of air into the smoke generator, through the smoke oven and out the chimney. This method is usually confined to smaller machines.

2. Forced and/or induced draught. When insufficient natural draught can be produced it is necessary to fit a fan. The size and rotational speed of this is selected to ensure that the necessary quantity of air can be forced through the smoke generator, smoke oven and its passages and out through the chimney. The air flow passages are usually provided with dampers to regulate the flow and also there are sometimes baffles to improve the distribution of air within the passages. When the fan is fitted between the smoke generator and oven, the fan is known as an 'inlet fan' and the process is 'forced draught'. If the fan is on the chimney side of the oven it is an 'outlet fan' and the process is 'induced draught'. In the former case, the air is 'pushed' through, whereas in the latter, it is 'pulled' through.

3. Balanced draught. For better distribution and more sophisticated control — usually in large capacity smokers — a combination of inlet and outlet fans can be used to supply smoke to the oven and extract it from the oven. This can produce a very even airflow and smoke distribution.

Fuel

Hardwood is the preferred source of the wood chips or sawdust. Softwoods should be avoided as they contain resin which gives a bitter flavour to food and releases substances which are harmful. Sawdust is the form in which wood, once alight, will smoulder with least air and is ideal for smoking because, as has been previously explained, the aim is to keep the temperature low. Increased air flow will make the sawdust or chips burn at a higher temperature and so reduce the ability to cold-smoke. When wood is not available substances such as rice husks, grated corn cobs and bagasse can be used, if possible mixed with a little sawdust. The size of the sawdust chips should be quite small, of the type produced by a hand saw or fine chain saw and each machine will have a recommended size best suited to it, depending on the flow of air into the smoke generator. Most commercial smokers buy the sawdust in bulk from a proven supplier, but individuals may wish to make their own. If you have access to a chain saw in the preparation of sawdust, it must be lubricated with vegetable oil to avoid mineral oil flavours in the food.

Ancillary equipment

On a commercial basis, your premises should be fitted out with gutting or **preparation tables** which, ideally, should be of stainless steel and have an overhead water supply with conduits to each operator. You will also need **brine tubs** of food-quality, heavy-duty plastic with drainage plugs, in a variety of sizes, to suit the products being handled. Tubs should be allocated to a particular product and not interchanged. If possible, the tubs should have wheels or be equipped with a dolly with wheels to avoid heavy lifting.

A **salt saturator** is the best way to ensure a supply of brine at the correct salinity, and avoids an operator having to mix salt and water. This is a water tank containing a pre-determined amount of salt and water and which is calibrated to deliver this brine solution to the workplace. It is therefore usually situated off the ground to allow ease of filling of brine tubs and to allow the solution to pass via a filter and tap at the bottom of the tank via a delivery hose. By delivering a strong brine which can then be diluted to the correct salinity, time is saved in mixing salt and water in each brine tub when required.

A **range of knives** are needed to do different jobs, such as gutting, boning, filleting and slicing. A **chain-mail** glove to protect the hands is a worthwhile investment.

If a salt saturator is not used, **paddles** made of high density plastic will be needed to mix brines in the tubs. A **brineometer** or **salinometer**, as it is sometimes called, will allow you to test brine strengths and to monitor losses during brining. This ensures that salt has been absorbed into the material, because if there is no weakening of the brine then there is some reason for the salt not being absorbed.

A **ham boiler** can be filled with water and maintained at a set temperature of 75°C/167°F for most products. Joints are then put into special cooking bags, then left to slowly cook in their own juice for a given time, which again varies according to the product. Around four and a half hours from cold immersion to an internal temperature of 72-80°C/162-176°F. The weight loss for this particular type of cooking is low.

A **weighing machine** for the salt and a **weighing/labelling system** for enabling the correct labelling of packaged goods will be necessary if you are despatching your own produce. If you are going to be despatching on a large scale, **a banding machine** for cartons saves a lot of time and looks more professional than the other methods of securing the lids onto the box. Some means of keeping the boxes cool is also required, and **ice flakes** are the most usual method. If refrigerated transport is used, the ice will remain in good condition, but sometimes a double container is required to ensure no leaks, such as may occur on air transport. **Dry ice** can be used but is expensive and must be declared to air freight forwarders or airlines due to the carbon dioxide fumes which are created as the temperature rises. **Cold blocks** which are used in **cool boxes** are ideal for short-haul journeys on a local basis, but are also manufactured to commercial specifications and are reuseable. Sheets of special material which after soaking become chilled, can also be used to good effect. In any event, use the system which best meets your requirement.

Waste scraps of food should be collected into **bowls** with clear segregation between usable scraps and others. The bowls can then be emptied frequently, either into refrigerated containers for re-use or into the refuse bin. This avoids scraps ending up in piles or on the floors and thus being carried into another part of the factory on footwear.

Chopping boards, if used, should be kept clearly marked for each product and thoroughly cleaned, not only at the end of the day, but frequently during the period of use. An **alcohol-based cleansing spray** which is then dried off with disposable paper towel is an easy and efficient way to keep working surfaces hygienic.

If you are involved with the slicing of your product, careful thought must be given to the choice of **slicing machine**. In the meat and pork industry there are many simple machines which will do the job very well for a limited initial outlay. Should you be required to handle salmon, the problem is more difficult. The slicers required are extremely expensive and most need the salmon side to be deeply chilled or almost frozen. If the product has already been frozen it is illegal to sell it without notification that it is 'unsuitable for freezing'. Pre-frozen products should never be refrozen after thawing because bacteria develop during thawing and can therefore contaminate the food further.

It must be said that at present this regulation is not widely observed. Slicing by hand is a matter of experience. It takes many man hours to justify the cost of a commercial slicing machine, which should only be purchased when the volume of output can justify it. Machines are obviously efficient but cannot always be adapted to different shapes of fish. The best compromise for commercial concerns is the round, electrically operated blade machine called the Wizard which was designed for de-hairing hides. It does rely on the skill and imagination of the operator, however, so only practice will make perfect.

There are many trade magazines offering help with this side of preparation, and it is a very good idea to thoroughly shop around before you make any decisions.

What of vacuum packing? If you are hoping to supply a large retail outlet you will need to vacuum pack because it gives the best protection available at present, but it is not ideal and certainly doesn't improve the taste of the product. In this area, as with slicing, look at as many systems as possible, and find the one most suitable to what you are intending to produce.

Once you are up and running it will become clear what mechanical aids would make the operation more efficient but do not set out with the idea that you must be as automated as possible. People need employment and it takes a long time for an operator's wages to come anywhere near to the cost of expensive filleting and gutting machines. With these machines there is a lot of waste because they must be set for a particular size of fish or they will not give good results. You need to have a large tonnage of one particular size going through or the time taken to set up the machine for different sizes of fish makes the production costs too high, not to mention capital outlay needed to buy the machine.

Remember, the simpler your equipment, the easier it is to keep clean and maintain.

Equipment and ingredients checklist

Select equipment that can be easily scalded. Stainless steel or heavy-duty food-quality plastics are very suitable. A high standard of cleanliness will ensure a low failure rate and this is easier if the equipment is straightforward, without corners and excess handles.

The following list of basic equipment will cover initial brining and smoking and can be added to as required:

Blowtorch

Needed to set the sawdust alight. Tamar and Ronson manufacture them.

Brine pump

A syringe with a perforated needle allowing brine to be injected into thick muscle or along the bone of cuts being cured over a long period of time. This ensures even distribution of brine and prevents spoilage. See illustration on page 32.

Brine tubs

This can be a box, dish, pan, tub or other container large and strong enough to hold sufficient brine to cover food completely. It is preferable to have a lid, but not essential. Wheeled-base tubs or a dolly for large tubs facilitates movement.

Fuel

This is normally any hardwood sawdust. With some smokers a mixture of sawdust and shavings will work. This is dependent on the amount of air available, because if there is too much air, the fuel will burn and not smoulder. All British hardwoods are suitable, and those from foreign countries where the resin content is low.

Hooks, rods and skewers

A means of supporting the food in the smoker other than on shelves. Hooks can be single or double, short shanked or long. Stainless steel rods are useful to thread fish along to enable greater loading in the smoker. Stainless steel triangular section rods can be used to hang two fish, tied at the tail, while smoking.

Gloves

A chain-mail glove for protection when gutting and filleting or when boning out is a good safety precaution. (Aprons of similar stainless steel links are also available).

Knives

A series of knives to cover all different operations are required and should be kept sharp. Knives should not have wooden handles as they are a hygiene risk.

Long-handled spoon

A long-handled spoon or paddle is needed to mix brine.

Meat thermometer

A thermometer with a probe about 7.5cm/3 inches long to enable the internal temperature of food being hot-smoked to be recorded. Available with hand-held digital read-out. See illustration on page 69.

Meat saw

A small saw with adjustable blade useful for cutting through bones when preparing joints.

Notebook/daybook

An essential part of the process to record brines, brining times and smoking times to ensure that a successful result can be repeated. Also needed to allocate batch numbers and date and destination of consignment.

Salineometer/brineometer

Instrument which gives a reading of the strength of salt solution as a percentage.

Salt

The essential ingredient of all brines.

Salt saturator

Tank installed to give constant supply of brine at a given percentage salinity.

Saltpetre

Potassium nitrate, used in small quantities to inhibit bacterial growth and to provide the chemical agent which stabilises the end-product which can then be described as 'cured'. Proprietary brands of curing salts containing agents are available on the market but analyses of the results should be studied to ensure that acceptable levels of nitrates or nitrites are not exceeded.

Weighing scales

To weigh and record food before and after brining to assess weight loss. Used to ascertain total weight and therefore amount of brine required.

Seasoning

Herbs, spice or a mixture of both added to brines to give a particular flavour. Juniper, mace, rosemary, cloves, bay leaves, allspice and peppercorns are among the most widely used.

Sugar

Added to brines to counteract the hardening effect of salt. Sugar can be white, soft brown, demerara, black Barbados or any other. Molasses or honey may also be used.

Useful spices

Spice	Flavour	Uses
Allspice	Fragrant, like a mixture of cinnamon, cloves and nutmeg.	Whole, in stock for poaching meat and fish. Marinades.
Aniseed	Faintly liquorice.	Rich meat dishes. Beef stews, casserole of hare or pigeon. Sausages.
Caraway	Slightly sharp and peppery.	Rich game, meat and offal dishes to offset fattiness. Good with pork and goose.
Cardamom	Slightly lemony and bitter-sweet.	Good with beef, pork, goose. Used to flavour pickled herrings.
Chillies	Hottest of all peppery spices.	In any recipe, calling for hot spice.
Cinnamon & cassia	Gentle, sweet, musky flavour.	Cinnamon stick in pickling, sprinkled on meat. Whole cassia buds used in curries.
Cloves	Sweet and tangy.	Whole in brines. Studded in ham and gammon. In pickle for pork and beef.
Coriander	Mild, sweet and pungent.	Spiced sausage and meat loaf. In brines.
Cumin	Strong and aromatic with lingering flavour.	In marinades or brines for lamb and beef.
Fenugreek	Slightly bitter and caramel-like.	Whole in stock. For poaching fish.
Ginger	Hot, rich flavour. Ground ginger less pungent than the root.	Whole in pickling. Ground sprinkled on grilled and baked white fish.

Spice	Flavour	Uses
Juniper	Sweet, aromatic and pine-like.	For brines and pickling. Good with all game, poultry and pork. All meat and fish marinades.
Mace	An exotic spice with strong nutmeg flavour.	Whole in picklings, sausages and casseroles. Ground in paté, minced beef and meat loaf.
Mixed Spice	Cinnamon, cloves, allspice, some ginger.	Rubbed into beef for pickling. Game casseroles and meat loaf.
Mustard Seed	Slightly sharp and hot flavour.	Whole in pickling meat and fish, especially herring and mackerel. Casserole of pork, veal, rabbit and offal.
Nutmeg	Exotic, sweet musky flavour.	To spice paté and meat loaf. Grated into fish. Sauces and white chicken dishes.
Paprika	Mild and slightly sweet, not at all hot.	With shellfish, especially prawns, shrimps. Meat dishes with tomato.
Pepper, black	Strong, pungent, spicy.	Whole peppercorns for pickling meat and fish. In marinades. Ground on all dark meat dishes.
Pepper, white	Milder and less exotic.	Whole peppercorns in pickled meat and fish. Ground in white fish and poultry dishes where black affects the appearance.
Pickling Spice		

Spice	Flavour	Uses
	Black and white whole peppercorns, chillies, mustard seed, cloves, allspice, ginger, mace, coriander. Flavour can be varied by ingredients and proportion. Mix your own.	To pickle beef and pork and souse or pickle fish.
Poppy Seed	Scant flavour, nutty texture.	Whole seed in brines.
Saffron	Exotic golden colour, slightly sweet taste.	Powerful colourant for fish and rice.
Sesame	Sweet, nutty flavour and crunchy texture.	Toasted to top creamed fish dishes and chicken.
Tamarind	Sour, acidy flavour.	Authentic sour taste in meat and fish curries. Sharper than lemon or lime juice.
Turmeric	Bright yellow colour, rather bitter.	Rice and fish dishes for colouring, cheaper than saffron.
Vanilla	Sweet, chocolate-like flavour.	

Herbs useful in smoking

These are useful, whether fresh or dried, as a useful addition to brines to impart additional flavour. It is always wise to experiment with a herb, or a combination of them, to ensure the end-product is pleasant. Some herbs go better with certain foods than others, and the following list may provide a starting point:

Herb	*Flavour*	*Uses*
Basil	Strong and powerful giving a sweet delicate pungency.	With oily fish, lamb, chicken duck, goose. Combined with tomato.
Bay	Aromatic slightly bitter. Stimulates the appetite.	Part of a bouquet garni. With oily fish, veal, sauces for pasta. Paté and terrines.
Chervil	Fragrant, slightly sweetish with an aromatic scent and taste.	Delicate fish and shellfish.
Dill	Sharply aromatic yet slightly sweet. Fragrant.	With Salmon. Fish soups and stews. Mutton.
Fennel	Slight flavour of aniseed. Fragrant	Oily Fish. Lamb and mutton. Sauces. Fish soups and stews.
Garlic	Strong-smelling and pungent taste.	In small amounts with almost anything.
Lemon Balm	Strong lemony scent.	Fish, lamb, chicken and other poultry.
Marjoram	Strong, sweet yet spicy.	Fish, beef, pork, sausages and liver.
Mint	Spearmint flavour.	Eels, fish and mutton.
Oregano	Strong spicy sweet	Fish soups, stews, fish, pork.
Parsley	Fragrant.	Everything.
Rosemary	Fragrant with pungent resinous taste.	Eel, halibut and other strong-tasting fish. Lamb.

Herb	Flavour	Uses
Sage	Pungent. One herb which must be carefully dried or the wrong flavour is obtained.	Poultry and venison.
Savory	Piquant; strong and slightly peppery.	In small amounts with eels and flat fish. Beef, pork, sausage and liver.
Tarragon	Strong, aromatic flavour which is unusual, being sweet but slightly bitter.	Fish. Poultry, venison.
Thyme	Pungent, clove-like.	Chicken, Fish, meat, hare and rabbit. Use with discretion. Fatty fish. eel, pork, sausage.

Chapter 6

Smoking Techniques

TO CALCULATE the length of time food must be exposed to smoke after brining is difficult to do because there are so many variables. These are:

1. Personal taste
2. Type of food
3. If the food has been processed before smoking, e.g. cheese, eggs (boiled first)
4. If the food does not require cooking, e.g. nuts, salt
5. If the food is to be eaten cold, or...
6. Stored and eaten cold, or...
7. Eaten hot, or...
8. Stored and eaten hot

All foods should be cold-smoked for a period in the temperature range 21°C-31°C/70°-88°F. If the food is then to be frozen and subsequently cooked before eating it is not necessary to hot-smoke. With all other combinations of storing and eating, hot-smoking must be done to ensure safety. The flavour develops during the cold-smoking period, and therefore, ordinary forms of cooking are acceptable as far as taste goes, but the hot-smoking produces an attractive darker colour on the outside.

The higher the fat content of the food being smoked, the longer it will have to be in smoke, because it is more difficult to dry and reach the required weight loss. The recipes given here are for various categories of food, and by keeping careful notes of brining times, strengths and the length of time in smoke, it is possible to successfully smoke products which will fit into one of the categories. Remember that the bulkier the food, the longer it will take for the smoke to penetrate the whole way through.

Cheese

No preparation in the form of brining is required because cheese has already been processed with salt added. To avoid cracking of the outside by over-drying, the cheese should be cut into slabs not in excess of 3.5–5cm/1½–2 inches thick and placed in the smoker. Cold-smoke two to four hours for a mild flavour and five to seven hours to produce a richer, stronger smoked flavour.

Choice of cheese is widespread among the red and white varieties, and Edam and other cheeses of that texture also smoke well. Avoid the more exotic varieties as this would appear to be gilding the lily, although they do smoke well. Goats and ewe's milk cheese are also possibilities, but extra care with the making of those cheeses is recommended to avoid bacterial problems. Cream cheeses need very much less time, and because of the extra moisture present can sometimes taste a little bitter. Cream cheeses such as Bonbel smoke very well once the wax coating is removed.

Eggs

Not a very well known smoked product but eggs provide a subtly different flavour for use in salads, sauces, and as the basis for egg mayonnaise or in the centre of a game pie. The eggs can be of any bird and should be fresh, although not newly-laid because the shells will not come off easily and in doing so may remove some of the white at the same time. Put the eggs into a pan with water at room temperature or if straight out of the refrigerator, cold water. Bring slowly to the boil. When a rapid boil has been reached, remove from the heat and leave them in the water for 15 minutes. Run under the cold tap to stop blackening of the yolk edge. Shell and put into the smoker. Cold-smoke for four to five hours, or until they turn a pale straw colour.

If a brine is available which has been used for anything other than fish, the eggs can be put into it for five to 15 minutes depending on size. Dry off and smoke as above.

Alternatively salt and pepper can be shaken onto the shelled eggs prior to smoking but this does give the finished product a mottled, dusty appearance.

Nuts

Almost all nuts will smoke, such as almonds, walnuts, brazil nuts, hazelnuts, pistachios, peanuts, sunflower seeds and pine kernels. They should be placed in shallow trays or on fine perforated mesh (pierced aluminium foil will serve). Sprinkle with salt and cold-smoke. Three to four hours is usually enough but if a roasted, smoked product is required, hot-smoke at 93°C/200°F for 15 minutes.

Salt

Shallow trays of salt, cold-smoked for three to four hours, acquire a golden colour and a lovely, smoky flavour which makes this useful for seasoning.

Fish

Fish can be divided up into a number of groups. There are sea fish, freshwater fish, white fish and oily fish. White fish, either from the sea or freshwater, smoke more quickly than the second category of richly fleshed fish.

Never be tempted to smoke any fish other than fresh fish or fully thawed-out pre-frozen fish, because any off-flavours will increase during the time in the smoker. If the fish has been frozen, thaw it out thoroughly and proceed as for fresh fish. Frozen fish flesh will be much softer than fresh and will absorb salt more quickly. Handle with great care to avoid marking or bruising the flesh. For frozen fish, especially fillets, it is better not to hang them but use mesh trays instead. When fish are strong enough to support their own weight, which is usually the case in fresh fish or whole, thawed-out pre-frozen fish, they can be suspended on hooks which are passed through the eyes or gills and then hung on stainless steel rods, or tied together in pairs at the tail and slung over the rods.

Some difficulty may be experienced with thawed fish which are found to be excessively soft. This is sometimes due to the fact that they have been taken from the sea too quickly and have suffered 'the bends'. They are not suitable to smoke.

The group of white fish best known in European waters includes haddock, Norway haddock, cod, whiting, pollok, coley, ling and hake. The same group is found in North American waters with the addition of cat fish, red snapper, redfish, Boston bluefish and black cod. This in no way exhausts the list, but will allow a type of fish for comparison with any other species. Among the richer fleshed fish we have Atlantic salmon, (Salmo Salar), and the group which is usually lumped together as Pacific salmon: chum, humpback, sockeye, coho and chinook but which all belong to the Oncorhynchus species rather than Salmo. Other oily-fleshed fish are halibut, seatrout, turbot, sturgeon, seabass, herring and mackerel. Smoked salmon has developed into a worldwide luxury food with halibut, sturgeon and some turbot being processed in the same way and thus providing some variety at the upper end of the smoked fish market.

Herring has been very widely used in Europe for smoking but only in Britain does the 'kipper' or cold-smoked herring appear. Europeans have tended to hot-smoke the herring, relying on cooking and packing in salt or oil to provide preservative qualities. The difficulties in the supply of raw herring due to overfishing are seriously affecting the herring smoking industry and in Northern Europe supplies of good quality, fatty herring from further afield are widely sought.

Mackerel, once smoked, has become a popular food and the variety in quality is legion. Many people claim that they are made unwell by eating smoked mackerel but it can usually be traced to a dye used in the process, rather than a direct fault attributable to the fish. Freezing whole mackerel is not advised, so ensure that the fish are gutted beforehand.

The following recipes cover groups of fish and their smoking in order to give guidance for any type of similar fish.

White fish

White fish are normally smoked as fillets for further processing in cooked dishes. They can also be smoked whole then stuffed and baked, providing an alternative to a joint of meat.

The flesh of white fish has a high water content so it is more satisfactory to use a strong brine for a short time then a weaker brine for a longer time. Dry-brining is a possibility but the short duration time involved does not make this a commercial proposition, except when dealing with very large fish. Today, the average weight of white fish being processed rarely exceeds 1.5–2.5kg/3lb 5oz–5lb 8oz gross, with the resultant salting fillets of about 0.5–1kg/1lb–2lb 3oz.

To prepare the brine mix 4.5litres/1 gallon of cold water with 1.22 kg/2 lb 10½oz of rock or sea salt. Ensure that the salt is completely dissolved before putting in the fish. Fillets of 100g/3½ oz or less will require only 5 minutes in the brine. 225g/8 oz fillets will require 10 minutes and 500g/1 lb fillets around 15 minutes. Whole fish require 30 minutes to a full hour.

Remove from the brine and allow the fish to drip-dry until the surface is tacky. To put wet fillets into the smoker will not only make the smoker damp, and consequently take longer to smoke the fish, but the surface of the fish will have patches of white, where the salt has dried off unevenly, instead of an overall gloss.

In very damp atmospheres where there is no dehumidification, a fan heater to help dry the air is useful and cuts down the air-drying time prior to smoking. Once the surface is tacky the fish should be cold-smoked for three to four hours for small fillets, five to six hours for large fillets, or for whole fish when opened out, as is the case with a Findon haddock. Whole fish that have not been filleted will require seven to eight hours. At the end of those given times the colour will be beginning to develop but will deepen after removal from the smoke. This is due to the continued interaction of the salt and the smoke.

Weight loss will be around 12–14%. Fish smoked in this way are normally cooked before eating.

Herring

Smoked herring are kippers. The fish should be thick and in good, fatty condition (June to October) to ensure a juicy kipper. Split the fish through the head and continue the cut down one side of the backbone. Spread the fish open and gut without damaging the belly wall. Detach the backbone from the head. Using a 70% brine, small fish are given three minutes, medium-sized five minutes and large good-quality fish will need 10–12 minutes, depending on thickness. Drip-dry for one hour. Finish is not so important as with white fish because the oil will give a natural gloss.

Smoke at 30°C/80°F or below for six to eight hours until golden in colour and possessing a good smoky smell. Weight loss will be between 14–18%.

In Europe herring are traditionally hot-smoked but without any prior salting. Due to the lack of cold-smoking, they have little flavour and taste like boiled herring. If cooked herring are required, a suitable procedure is as follows:

Gut the herring from vent, or anus, up to the throat. Leave the head on because it is easier to hang the fish. Wash thoroughly. Using an 80% brine solution, steep the fish for 10 minutes and then allow to drip-dry. Cold-smoke overnight and then raise the temperature slowly until the dorsal fin comes away easily, which should take about 45 minutes.

Recipes for herring can be used for fish of a similar character.

Buckling

Buckling are not well-known in Britain. They are whole herring, with head, guts and roe left in and the whole fish is smoked. Wash the fish and brine for three hours in an 80% solution. Wash in cold water and drip-dry. Hang the fish on rods and cold-smoke for two hours. Raise the temperature of the smoker to 40°C/104°F and finish cooking until the dorsal fin pulls out easily. Weight loss will be between 20–25%.

Sprats and whitebait

Make a small cut below the gills and press entrails out through this cut. Rinse well in cold water. Brine for eight minutes in 80% solution. Drip-dry for 30 minutes and cold-smoke for four to five hours. Increase temperature to 90°C/194°F for 30 minutes or until cooked.

Alternatively, the fish can be removed after cold-smoking and grilled or deep-fried in order to be served hot.

Red herring

Herring, sprats or pilchards are suitable for making this heavily smoked type of fish.

Cut and remove the heads, but do not split open the fish. Some of the gut will be removed with the head. Dry salt for seven to eight days depending on the size of the fish. The amount of salt used should be half the weight of the fish and should be applied in layers of salt, fish, salt etc. Remove the surface salt by soaking in fresh-water for about an hour.

Cold-smoke until the fish are firm and dry. The time for this will vary depending on the humidity of the air in the smoking oven, the size of the fish and the oil content. Red herring are an unusual 'starter' and often an acquired taste.

Smokies

The fish called 'smokies' are associated with Arbroath in Scotland because only there are they made correctly. The small haddock from which they are traditionally made are cleaned without opening out the fish, then brined in 80% solution for about 30 minutes, tied together in pairs by the tail and suspended over a pit in which there are burning logs. The fire is doused with water and the fish covered over with hessian or other material, and the resultant particulate-laden steam cooks the fish and gives the outside a smoked appearance. This means of preparation is similar to some continental methods, such as the Bornholm herring, but cannot be termed smoked in the accepted way. (The island of Bornholm in the Baltic produces herring which are cooked in special kilns in which the rows of fish are moved back and forth over smouldering embers. They are then packed in barrels with salt to preserve them, or put in small wooden boxes for the tourist trade. However, because no salt is used before exposure to the smoke, there is no enhancement of flavour and no colouration).

To prepare smokies in a smoker prepare the fish as described, cold-smoke for three hours, put a basin (or basins) of water on the drip tray of the smoker, or on the floor if the smoke generator is separate, and raise the temperature to 90°C/194°F until the flesh is cooked. Not as good but tasty all the same!

Mackerel

To have the best result mackerel should be newly caught, and as soon as rigor is complete the brining should take place. If you are unable to catch the fish locally you must be very careful that they are fresh. A mackerel that bends when held horizontally is too old. Most smoked mackerel in the shops today is processed under air pressure and it is just not possible to compete with this product on a commercial basis when smoking properly. If, however, you are processing for a small market you will be able to sell all you can produce.

The fish are gutted, and the main artery which runs down the backbone is removed, then they are washed in cold water. Leave the heads on. Use a 70–75% brine and leave in for one and a half to two hours depending on size. Rinse the fish in cold water and string them on rods through the eyes. Drip-dry for two to three hours. Small sticks inserted to keep the belly cavity open and allow good smoke penetration are helpful, but not essential. Cold-smoke for four to six hours until the skin is dry and golden in colour. Raise the temperature to 82°C/180°F until the fish are cooked. Weight loss should be about 25%. The skin of the belly and flanks will turn a deep golden colour with fine wrinkles as they cool.

Trout

This recipe is for freshwater trout which must be hot-smoked because there will always be the possibility of botulism spores being present. This will give good colour. The trout are gutted and thoroughly washed in cold water. Fish of up to a 500g/1 lb should be brined in an 80% solution for half an hour. Larger fish should be left longer in proportion to their weight. After brining, allow to drip-dry until the surface is tacky. Cold-smoke overnight or for a minimum of seven hours. Raise temperature to 82°C/180°F slowly and remove when cooked. The dorsal fin can be easily removed when the trout are ready.

Allow to cool before packing.

If freshwater trout fillets are being smoked from opened-out fish, allow 30 minutes in brine, six hours cold-smoking and cook until ready at the same temperature as for whole fish. It is also possible to cold-smoke and hot-smoke whole fish and then remove the two sides of fish off the bone, giving attractive fillets. The skin should be removed once the fillets are taken off the bone because it holds the flesh together more easily.

Eels

It must be stressed that eels should be held in clean water for about two days to ensure that there is no muddy taste to the flesh. If the eels are farmed, a starvation period of at least three days is recommended. Care must be taken when gutting that the kidney is removed, or taint from the acid in that organ will be transferred to the flesh. The kidney lies below the vent.

Eels can vary in size from large metre-long congers to half-metre-long silver eels. Silver eels are found in rivers and ponds; they have a silver-toned belly and the flesh is not too coarse. The smoking process is the same except that the length of time in brine should be extended for congers. Eels are notoriously difficult to kill, and the quickest and most humane method is to make a salt solution with tepid water until no more salt can be dissolved. The container for this should either be very deep so that the eels cannot jump out, or be fitted with a tight-fitting lid. Place the eels in the container. Activity will be extreme for a second or so, then the movement diminishes. Leave them in this solution for 10 minutes. Remove and rinsethe eels in cold water. Having killed the eels, no more can be done until the onset and completion of rigor when brining can commence.

When the eels are ready, place them in an 80% brine solution and leave them for 10–15 minutes for silvers and 30 minutes for congers. Because of their high oil content, eels can be put into the smoker wet, but this is introducing moisture into the smoker, so if time allows, drip-dry until the surface is tacky.

Small wooden skewers can be put in the belly to hold the sides apart but this is not critical because the smoke will penetrate anyway. Cold-smoke for six hours, raise the temperature to 32°C/90°F for a further hour and finish them off by raising the temperature slowly to 80°C/176°F and maintain this until the eels are tender. Eels should have a good, smoky flavour with the flesh having a firm, buttery consistency. It should not be rubbery nor taste of too much salt. The weight loss of eels from the ungutted state to smoked is around 40%.

Grey mullet

I have singled out this fish because the roe are used to make botargo (see also under Fish Eggs on page 64) and therefore, once the roe is extracted, the fish smokes well. The scales are very tough and have to be removed before smoking. Use an 80% brine and leave the fish in for 20–30 minutes, depending on size. Allow to drip-dry.

Cold-smoke overnight or for at least 12 hours until straw-coloured. Red mullet can be treated in the same way.

Monkfish tails

The monkfish is no longer a discarded fish, and the tail and fillets from the side of the belly are a delicacy and are excellent when smoked. Use an 80% brine solution giving five to ten minutes for fillets, depending on thickness and 10–15 minutes for tails, again according to size. Cold-smoke until straw-coloured.

Skate wings

Skate wings should be brined in an 80% solution for five minutes upwards according to the size which can sometimes be very large. Cold-smoke until pale straw-coloured. This will require a minimum of four hours for small wings, increasing the time with the thickness of the wings.

Swordfish

Swordfish in Britain is usually available in steaks, but by far the best way to smoke this flavoursome fish is to take off hams from the bone. This gives a better shape which lends itself to different dishes, and it gets rid of the staining from blood vessels. Use dry salt or 70% brine and leave in for 10 minutes and upwards depending on thickness. Cold-smoke for six to eight hours. If there is a tendency for the flesh to dry, brush lightly with vegetable oil during smoking.

Sturgeon

See Salmon.

Salmon

The quality of the raw materials with which you start is vital when smoking this fish. Salmon are normally smoked as separate sides although if hanging the fish in the smoker, 'butterflying' is acceptable. This involves making a cut from the nose of the fish, through the head and down *one* side of the backbone, without cutting the belly skin. The fish is then opened out and gutted. The backbone can then be removed also, but it is easier to leave it intact to help support the flesh during smoking. The salmon are then brined and threaded on to rods under the lug bone and hung in the smoker. In some farmed fish, or fish which have been frozen for a long time, the bones are not strong enough to support the flesh and butterflying is not recommended.

The salmon is gutted, cutting from the vent up to the head and all the intestines are then removed. If possible, always use bled fish, but, if this has not been possible, try to remove the gut and organs in one piece without piercing the heart or kidney. By minimising the amount of blood spilt, the fish are kept clean and there is little or no chance of infection from bacteria. Scrape the back bone clean and wash the gutted fish thoroughly in cold water. If two fillets are required, keep the head on as this gives you something to hold on to while filleting. To assist salt penetration, patches of skin are removed at the head and tail ends and in the middle of the back on both sides of the fish. This is best done before filleting. The reason for this is to enable salt to better penetrate the fatty layer just below the skin. If this is not done, the time taken for normal salt penetration will be such that the finished fish will be too salty.

If you are selling on the fish, always check with your customer the percentage salt he requires because this determines the length of time you must leave the fish in the brine. When doing your testing, have the salt content analysed by a professional food analyst and repeat this from time to time. There is a mistaken idea that all salmon are the same, but this is not so. Often the results of these tests will be an indication that all is not well with the fish you are buying. If, for instance, the expected salt level is not reached, it means that the correct degree of osmosis has not taken place and the reasons for this have been mentioned in Chapter 3, Selection of Raw Materials. When exporting smoked salmon, an irregular reading of salt content may result in the return of a whole consignment. The importance of knowing the source and treatment of the salmon you buy is essential, for instance, are the fish packed before the onset of rigor? If so, they must be removed from the packing ice and rigor should be allowed to take place on delivery to your premises. This is extremely important, and must not be ignored since it is central to the quality of the end-product and therefore your entire business.

There is really no precise formula for the successful smoking of salmon because all fish are different, and only experience and interest in the quality of the end-product will give consistently good results. There are, however, certain factors which will influence the time in brine and smoke.

1. The individual preference for smoke flavour, the consistency of flesh colour, the time taken between smoking and eating and the desired specification from the customer. These factors alone can

produce brining times which vary from six to 30 hours, or longer. Smoked salmon as an end-product can be made to vary significantly. Lox, the name given to lightly salted, smoked salmon popular in Jewish communities and in New York, requires only six hours brining and 10 in the smoker. The traditional Scottish smoked salmon is dry-salted overnight or brined for a similar time and smoked for about a day in the high smokehouses rarely seen today. North American Indians dry-brine the fish for even longer and smoke it until it is almost black. The range is therefore great, and that is why it is a good idea to establish the type of finished product your customer has in mind.

If the salmon is going to be eaten as soon as it is smoked, the salt content can be kept to around 2.5%, but if required to have a shelf-life of the three weeks or so upon which some supermarkets insist, then 3% is more in line. This makes the salmon saltier unless a longer smoking time is given to enable the enzymatic action to change the chemical make-up of the flesh.

If the fish are to be chilled but not frozen, then around 2.5% is acceptable but the supermarkets prefer a higher content as a safeguard.

It is well to remember that flavour usually denotes a better production method and consequently, a longer shelf-life. A salty, faintly smoke-flavoured product usually is produced by a quick method of smoking in pressurised smokers relying on the salt to give shelf-life, whereas the longer, slow smoking will allow the combination of acetates and aldehydes with the salt to bring about a better end-product.

2. The type of salmon, for example wild or farmed Atlantic (whether fresh or frozen), imported Atlantic, Pacific salmon, or salmon produced on farms in other parts of the world from Atlantic salmon ova. It is always wise to remember that Pacific salmon are not true salmon, but relatives of the cod family and thus subject to the possibility of worm infestation when in warmer waters. If you are exporting to the United States, the regulations issued by the Food & Drugs Administration should be checked to ensure difficulties with embargoes are not encountered with these type of fish.

3. The moisture content in salmon after brining, and the moisture content required after smoking. Too much oil in the raw material can upset both these readings.

4. The humidity of the air and smoke during smoking. Moisture will be removed in the oven by allowing smoke-borne vapours to combine with the salt which, in time, has a drying effect. To speed up the process some smokers rely on high velocity airflow which dries the fish, but as the air is moved past the fish too quickly, the vapours never settle and cannot combine with the salt. There is thus

no reaction with the flesh and no smoking occurs. To counteract this, a fine spray of water is sometimes used to keep up the moisture content in the smoker, but this is apt to mix with the carbon content and give a more tarry deposit.

5. The percentage salinity of brine or dry salt, and the time taken to salt the fish.

6. The weight and thickness of the salmon side and its oil content. The oil content of wild salmon will depend on the time of year, whether it is captured before or after spawning. This will also have an influence on farmed salmon, although their natural cycle is different, as will the type of feed used, the temperature and movement of the water and the stocking density.

Bearing these points in mind, we can prepare to smoke salmon. Dry-salting has been the traditional Scottish way of salting, and it is still a good method. It takes a little more experience to know what time to give a fish and when it is salted enough. A guide would be to take the weight of the side of salmon to be smoked, take one third of that weight in salt and use this amount to salt the side. Make a bed of salt in the bottom of the container or on slabs and lay a side, skin side down on top of this, then sprinkle salt on the fillet, lay down another side, skin side down, and so on until the sides are all sandwiched with a layer of salt. Allow to salt all night, and in the morning make a 40% brine solution, and wash off the surface salt by leaving the fish in this solution for 30 minutes. Remove from the solution and allow to drip-dry until the surface of the fish is tacky.

For beginners, however, it is easier to use a brine solution to salt the fish because they will come out more evenly salted if left for a sufficient time in a brine of medium salinity. The first method is to put the sides in a 60% brine solution, to which you can add a little brown sugar to prevent the hardening effect of the salt, and leave the sides in this all night. When the brining time is over, remove from the solution, allow to drip-dry and proceed as for dry salting. Alternatively, an 80% brine solution can be used, and the fish left for two and a half to four hours, depending on the size of the fillet. It takes salt approximately one hour to penetrate one quarter inch/6mm of salmon flesh. When a fish is salted uniformly the texture is springy, not hard, and should be the same the whole way through the fish.

If you are a beginner, you should always weigh before brining and after brining and drip-drying. With dry-salting, weight loss will be around 8%. When brining it should be 2% or a little more. If no weight loss is obtained there is something wrong with the fish. After air-drying, cold-smoke for a minimum of 10 hours for the smallest sides and increase the time until the sides have a golden glow, are flexible when bent, and when you run your finger down the side, an oil trace is visible, not water. Remove from the smoker and allow to mature for at least 12 hours before boning. The pin bones run down the length of the fish and should be teased up at their thin ends, prised back gently, then removed at the join with the backbone with sterilised tweezers or pliers. The lug bone, which is the shoulder bone at the top of the side and is kept on until after the process to keep the side supported while handling, should also be removed. The next step is.slicing and packing. Maturing should be undertaken in a chilled atmosphere at 3°C/38°F. It should be said that a large good quality salmon will require at least 18 hours in cold smoke, because the chemical changes brought about by the enzymes start on the outside and slowly work through the side causing the change in colour and flavour.

To recap, this is my step-by-step summary of salmon smoking for beginners.

While the fish are in rigor, the brine can be prepared or dry salt weighed.

1. Gut the fish
2. Remove patches of skin
3. Fillet
4. Brine or dry-salt
5. Drip and air-dry
6. Smoke for a minimum of 10 hours
7. Allow to mature
8. Remove pin bones and cut off lug bone
9. Trim and slice (if required)
10. Pack and label (if required)

Halibut

Gut fish, remove fillets and brine overnight in a 50% solution. Smoke as for salmon. (Slices cut across the width of the fish make good starters).

Turbot

See Halibut.

Sea trout

As for salmon for a cold-smoked product, but if required as a cooked, finished product follow the instructions for freshwater trout.

Fish eggs

The eggs of fish are held together in an egg sac. Collectively eggs are known as roes and as milt for the male eggs. These eggs can be cold-smoked and the product varies from ripe eggs stripped from a mature female fish and smoked on a suitable surface as individual eggs, to the whole egg sac roes. When smoking mature eggs the membrane will break very easily and a muslin-covered tray should be prepared to receive the eggs. Lower this carefully into a tub containing 80% brine solution. Do not fully immerse the shelf or the eggs will float off. Allow one minute and take the shelf out. Cold-smoke for three to four hours. Put the eggs into jars and cover with an air-tight lid. Keep refrigerated, and eat as soon as possible for maximum flavour. (Sea urchin roe can be smoked, but in order to have enough quantity to be viable, a large number are required. As a personal experiment remove roes, brine for two minutes in 70% solution, and cold-smoke for three to four hours).

Whole roes, being larger and less delicate, are easier to handle, but because the membrane surrounding the roe is delicate, minimum handling is required. Prepare the shelves of the smoker by covering with secured butter muslin. Place the roes on the prepared shelf and lower this into a 70% brine solution. Leave large roe for an hour, smaller roe for 30 minutes.

Remove from the brine and allow to drip-dry. There are two opinions — the first is that roe should be smoked until the outside is dark red/brown and quite dry. The second is that the roe should be smoked only long enough to impart flavour so that the outside can still be easily removed, and the roe made into sausage-shaped units for slicing. For the first option, smoke for at least 24 hours with a weight loss of around 30%, or cold-smoke for seven to eight hours, raising the temperature by 10°C for the last 45 minutes.

Using whole roes, usually of smaller sized haddock, cod, mullet or coley, a delicacy can be produced which is thinly sliced and served as a starter. In Mediterranean countries, these roes are also dried and known as botargo. The method used is the same for both smoked and dried roes up to the point of smoking. Roes are not uniform in size, and can be difficult to serve once finished. To overcome this the use of large sausage skins can be helpful. Use only natural skins, or a material which will allow the smoke and/or air to penetrate. Avoid plastic at all costs.

Carefully pack the roes, still in their membrane, into the skin and secure at one end. Hang this by the other end and this allows the weight of the eggs to compact their mass. Mix a 60% brine solution, and soak roes with a diameter of 5cm/2 inches for one hour, and 30 minutes for smaller diameter roes. At this stage, decide if you want the added flavour of smoke when air-drying. If so, cold-smoke for six to seven hours and then allow to air-dry in the smoker without lit fuel boxes for at least 12 hours, keeping the chimney damper fully open. If no smoke has been applied, allow the roes to air-dry for a full 24 hours minimum. Hang the roe sausages up in a cool temperature for a further 24 hours to mature and lose further moisture.

Tuna

Although instructions have already been given to prepare and smoke the mackerel found around European waters, the popularity of tuna (which is related to the mackerel) requires some comment. Among the more usual are skipjack, bluefin, wahoo, yellowfin and albacore, and each of these can be dealt with in the same way. For the best results, the fish have to be treated with care from the point of capture to delivery to the processing premises. Ideally, they should be bled and skinned immediately, and the gut removed. Boxing without bruising would be ideal, but because the fish are not identified as going to a particular market, little care is normally taken. It is easier to pack a frozen steak than a whole fish and, consequently, this is how the greater part of the tuna catch is available. To smoke for maximum appeal and flavour the whole ham or section should be removed from the fish in much the same way as a salmon fillet is removed. Brining is also done in the same way as for salmon giving a choice of an 80% brine for a short period of time, or overnight in 60% brine. Allow to drip-dry and smoke for eight to ten hours for sections weighing over 1.5kg/3lb 5oz. If being processed as steaks allow one hour in 75% brine and cold-smoke for four to six hours.

It has to be stressed that, if possible, source your raw material and insist on the killing and handling you want. This may be impossible due to the fact that most tuna are imported, but if you explain your requirements to the wholesale dealer from whom you buy, on their next visit overseas they can try to improve the situation for you.

Shark

All that has been said above applies to shark, and in most cases, shark are landed along with fish and are given little care. Some of the smaller varieties are caught in European waters and provide something unusual. Steaks, once again, are the more usual cut available and they can be brined as for tuna or dry-salted. Select steaks about 3cm/1¼ inch thick, 7.5 x 7.5 cm/3 x 3 inches in area. Sprinkle dry, medium, crystal, rock salt mixed with a little brown sugar on a tray or tub, place the steaks on the salt and cover these with more salt mixture. Leave for one hour. Rinse, air-dry for one to two hours, and cold-smoke for 5–6 hours, or until golden coloured. The steaks will then require some hot-smoking or other cooking to make them tender enough to eat.

Squid and octopus

In European waters there is a great variety of squid which is neglected as a source of food. The fresher the squid the more tender it will be, and as with all killing, it should be done with minimum stress. The method of battering freshly killed octopus to ensure tenderness cannot claim to be without stress. The body of squid is the only part used and can be split up to open it out rather than cut into rounds. Only the tentacles of the octopus are used and the body discarded, although the ink is prized as a delicacy in some countries. For both squid and octopus, use an 80% brine solution and leave squid for five to ten minutes (depending on the thickness of the flesh), and octopus for 10–20 minutes on the same basis. Cold-smoke for six hours, occasionally brushing lightly with vegetable oil to prevent the flesh from drying out. For very small squid, four hours is sufficient.

Shellfish

The increase in the number of farms producing shellfish means there are better supplies and consequently, alternative ways of offering this product to the consumer. Shellfish include scallops, mussels, oysters, clams, cockles and razor fish. Crabs, lobster, shrimps, prawns, freshwater prawns, are more correctly called crustaceans. All are delicately fleshed, and must be smoked in such a way to avoid drying out or coarsening the texture. Normal good husbandry will ensure that only properly prepared product will be harvested, but a word of warning on large freshwater prawns which are imported. Heat, at some stage in the process, is recommended, as there is no guarantee that the waters in which the prawns were caught have been monitored to record pollution. Where heat is required to kill the shellfish and open shells, keep this to a minimum by introducing them to hot water (or steam from hot water) which is at a temperature just sufficient to open their shells. Once open, remove from the heat source. Never use shellfish with open shells, since this means the fish are dead. Oysters are opened with a sharp, pointed knife which cuts the muscle joining the two halves of the shell together. Scallops are sometimes treated in the same way, but are more usually opened with hot water or steam as described.

Use a 40% brine solution and restrict time from one to five minutes for large scallops with the coral (the pink appendage) attached.

Always drip-dry before smoking. A quick brush with vegetable oil before placing in the smoker will ensure minimum hardening. Time in cold smoke will, as usual, depend on thickness, and should be from three hours for shrimps, oyster and small scallops, to five hours for large scallops. The colour of shellfish deepens quite considerably after removal from the smoker, so do not leave them too long.

It is possible to smoke various species in their half-shells, but this is difficult because only by removing the flesh and then carefully scrubbing and boiling the shell will a hygienic end-product be ensured. However, it does offer an attractive way of packaging and might be worth considering.

Meat

Smoked meat is not so widely known in the UK, but is becoming more popular as people return from overseas holidays having experienced the product. As with all raw material, meat must be mature enough to allow osmosis to take place and must be free of additives such as hormones or antibiotics. Hanging until tender should be no less than seven days. Smoked meat is not only offered as a 'starter' but can be the main item of a meal, or provide endless types of cold meat for slicing. The flesh retains moisture when properly smoked, and can be given distinctive flavours by the use of herbs or spices in a basic pickle. Lean cuts are preferable, such as topside, rump, brisket and fillet. Cuts of any weight can be cured and smoked, but when very large joints are being used, the use of the brine pump will ensure the core of the cut does not begin to spoil before the cure has penetrated from the outside. Dry salt can also be used, and providing the temperature is kept in the 3–4°C/38–40°F region, good results should be obtained. When possible, allow one or two days after curing before smoking, as this allows a settling of the cure and makes the flesh more receptive to the smoke.

The options explained in the smoking of venison apply equally to other meats and it should always be considered that customers may prefer to do the cooking of the joint after smoking and just before serving to obtain the best texture and flavour. Alternatively, customers sometimes prefer to use smoked meats as the basis of a salad or other cold dishes.

Basic sweet pickle

18 litres/4 gallons of water
1.2 kg/2lb 8oz salt
225g/8 oz sugar
1 tsp saltpetre
60g/2 oz pickling spice or flavouring of your own choice

The time required in the cure or pickle will be one and a half days per 500g/1lb weight. If the pieces of meat are long, such as a fillet or strips taken from the rump, the time should be cut by half. Take the meat out of the cure, rinse and hang up to dry overnight in an airy place or place in chill at 3°C/38°F. If possible allow to hang for two to three days before cold-smoking, as this allows the cure to even out and give less chance of too much saltiness in one area. Allow one day in smoke for thin, long cuts; two days for cuts up to 1.5 kg/3 lbs and three days for joints over 5 kg/11 lbs.

Cold-smoke for the length of time required to give a good smoky flavour throughout the cut of meat. Should any sign of overdrying appear, brush the meat very lightly with vegetable oil. Remove from the smoker and allow to mature for a minimum of two days. If time allows, increase this to one week. The production of smoked meat with a texture such as Parma ham entails hanging for nine months to a year, but not many producers can afford this.

I suggest that you always keep one piece of meat back from each batch and allow it to be kept for at least four months; this will then have maximum flavour and texture and will meet the precise requirements of at least one of your customers. Before the final storing to mature, wrap the meat in butter muslin and ensure that a cool temperature of 3°C/38°F is present in the storage area. This inhibits any bacterial activity. The cuts of meat should be wrapped in butter muslin, tied securely with string and hung out of direct sunlight in a current of cool air.

If the meat is going to be cooked after smoking, whether in the smoker itself or by another means, then the whole process is quicker. Use a 40% brine solution and add two tablespoons of brown sugar. Allow cuts up to 4.5kg/10lbs to remain for 24 hours in this and increase the length of time in proportion to weight, so that a 6.75kg/15lb joint would require 36 hours. Keep the brining cut in a chilled atmosphere. Remove and allow to drip-dry. It can be useful to cold-smoke, and then rub a seasoning to the outside of the cut of meat before hot-smoking. The undernoted seasoning is useful and can be kept in an airtight jar once mixed for easy use.

Meat seasoning

> 1.35kg/3 lbs salt
> 8 tbsp white or brown sugar
> 2 tbsp onion salt
> 4 tbsp celery salt
> 2 tbsp garlic salt
> 4 tbsp paprika pepper
> 8 tbsp white pepper
> 4 tbsp crushed dill (if the flavour is to your liking)

A digital read-out meat thermometer

Rub in the seasoning once the meat has been cold-smoked. If difficulty is encountered in making the mixture stick, a little oil can be added to the seasoning. Once the seasoning is added, and if cooking by hot-smoking, replace the meat in the smoker and raise the temperature slowly to 82°C/180°F. Allow 20 minutes per 500g/1 lb for rare meat; 25 minutes for medium and 30 minutes for well done.

A meat thermometer which can be inserted into the meat is useful because it gives an accurate internal reading and prevents under or overcooking. The meat is deemed ready when this temperature is reached, that is, 60°C/140°F for rare, 72°C/160°F for medium and 76°C/170°F for well done. If preparing meat for resale, 83°C/180°F is advised.

Steaks and hamburgers

A smoked steak is very succulent and can be prepared by cold-smoking for a short time and then cooking by hot-smoking, grilling or barbecuing. Use the most tender cuts, trimmed of fat for the best results. Commence by pounding a quantity of the basic seasoning or plain salt and sugar into both sides of the meat with a tenderising hammer. Leave for 15 minutes. Brush well with vegetable oil and cold-smoke for 30 minutes. If cooking in the smoker, remove after cold-smoking and raise the temperature up to 98°C/220°F, replace in the smoker with both sawdust boxes alight and leave till cooked the way you prefer.

For hamburgers minced beef can be rubbed with basic seasoning, mixed with a little oil then spread out in a shallow tray and cold-smoked for three hours. Stir the mixture around every 30 minutes. This gives a mild, smoky flavour to the meat which can also be used to make sauces or included in paté or sausages.

Beef ham

This is a traditional Scottish recipe.
10kg/22 lb rump steak (English topside)
1 kg/2 lb 3oz salt
250g/9 oz lb coarse raw sugar
150g/5oz saltpetre
15g/½ oz cloves
15g/½ oz allspice
30g/1 oz black pepper

Mix all the ingredients together thoroughly. Rub this over the surface of the beef and stuff as much as possible into the bone. (A modern variation to this might be the use of a brine pump down the length of the bone or use a piece of meat which is boned).

Let the meat lie for two to three days. Add another 500g/1 lb of salt, rub it in well, and turn the ham every other day. It will be ready in three weeks.

Wash and drain the ham and allow it to hang, or cold-smoke for one to two days. Allow the ham to air-dry by wrapping in muslin and hanging it in a cool, dry, dark place. Use it as you would bacon for frying, or bake it or boil it as preferred. This meat often forms the outside of beef olives. If it is too salty, soak the ham in cold water.

Lamb and mutton

Lamb and mutton are easily cured, but the cuts, once smoked, tend to become strong in flavour and dry out if not frozen or eaten quite soon.

Beer brine for lamb and mutton joints

2.25 litres/4 pints of water
500g/1 lb sea salt
240g/8 oz rock salt
1 level tsp saltpetre
240g/8 oz dark brown sugar
1 tsp freshly ground black pepper
1 tsp freshly ground allspice

Bring all the ingredients to the boil in a large pan and allow to cool overnight. Strain carefully. Rub the lamb or mutton with 55g/2 oz dark brown sugar mixed with 28g/1 oz saltpetre. Leave overnight.

Next day, rub the lamb again with the mixture. Put into a deep dish and pour the beer brine over it. Leave to cure for three weeks, but every day turn the lamb over and rub the brine in well, particularly near the bone. Keep in a temperature of 3°C/38°F. The lamb should then be cold-smoked for 24–36 hours or until desired degree of smokiness is obtained. Allow to mature for a further week before slicing. Alternatively, slowly bake the meat in the oven at 148°C/300°F for 25 minutes per 1kg/2 lb 3oz and serve hot. If you have access to a commercial ham boiler, this will cook very satisfactorily in one. The beef dry cure may also be used to cure lamb and mutton.

When a cured meat is not required the cut is placed in a 50% brine solution into which two tablespoons of sugar have been added. It is left overnight. After drip-drying, the cut is cold-smoked for 24 hours and roasted or cooked in the normal way without hot-smoking. This is a good recipe when using a best end or middle cut, but can equally well be used for a gigot or rolled shoulder.

Pork and wild boar

Because ham and bacon are made from pork, it is more usually associated with smoking than most types of flesh. It is lean in character and is cured easily. There are endless ways of using smoked pork as well as bacon and ham. The difference between them is that smoked pork need not be cured, but ham and bacon require a length of time in brine to cure.

The basic cure for bacon and ham is a 40% brine solution with 28g/1 oz saltpetre and two tablespoons of soft brown sugar. Allow two weeks for legs and 10 days for middle cuts. Drip-dry and allow to mature for a further week. Smoke for two to three days depending on preferred flavour.

Smoked pork fillet

The fillets of most pigs today are quite small and to produce the smoked pork fillet beloved of the continentals, we use the eye of the loin where the supporting bones and skirting meat have been removed and we are left with a uniformly round piece of meat. This will vary in diameter from two to four inches (50–100mm) and for our purposes, the larger size is the better. Brining times are one day per 5 cm/2 inch, and half a day extra. Because the fillet is round, the brine enters evenly and therefore one day is sufficient plus the extra half day. Certain kinds of pork will take salt in more quickly than others, so if the cured fillet is too salty just soak the excess out.

Pork fillet cure

40% brine solution
28g/1 oz saltpetre per 4.5 litres/1 gallon of water
55g/2 oz of sugar (any type will do)
At the end of the brining period allow the fillet to drip-dry and mature for one day before cold-smoking. 18–24 hours will be sufficient, and in order to stop the outside of the meat becoming too hard, gently rub it with some vegetable oil. The longer the maturing time after smoking the better, and if you wait a week and give it a further cold-smoke, this will intensify the flavour.

Wiltshire hams

Use a 5 kg/11 lb piece of boned leg pork. Make an 80% brine solution and add one tablespoon of saltpetre. Bring to the boil and ensure that all the salt is dissolved and the solution is clear. Skim the brine and cool. Put the ham in a suitable container and cover with the brine. Leave for one to two hours. Rinse. Make another brine solution from the following:
680g/1 lb 8 oz coarse salt
14g/½ oz saltpetre
1 litre/2 pints beer
225g/8 oz treacle molasses
12 juniper berries
14g/½ oz crushed peppercorns
Pour this over the pork. Cover and leave for two days per 500g/1 lb weight of pork at 3°C/38°F. Turn the pork every other day. Remove the pork from the beer solution. Rinse, dry and allow it to hang for a further 24 hours. Cold-smoke for 24 hours. Wrap in muslin and store in a cool, dry place out of sunlight for a further four to five days minimum. The longer the maturing time the better.

Canadian bacon

Traditionally made from the loin and the fat, back strip of the pig. Bacon is cured by brining or dry-salting in the same way as ham. Bacon, being usually a thinner cut and therefore requires less time in the cure, approximately one day per 500g/1 lb. Dry cure should then be applied at the rate of half the weight of the pork in salt, mixed with 25g/1 oz sugar and 1 teaspoon of saltpetre per 500g/1 lb. Half the amount of salt and sugar mix should be applied and rubbed into the pork and the other half applied three days later. Alternatively a sweet cure brine could then be used as a 40% solution with 225g/8 oz honey sugar or molasses and one teaspoon saltpetre added per 4.5litres/1 gallon. The pork should be turned over every two days and the brine stirred. When the curing time is finished the pork is removed from the solution or dry salt, washed and allowed to dry thoroughly before cold-smoking. 36 hours will give a mild, smoke flavour and for a more pronounced smoke flavour, leave the meat to smoke for a further 24 hours.

Spare ribs

Pork spare ribs are a tasty snack and are easy to smoke. Brining in 60% solution for 20 minutes or using a half salt, half sugar mixture rubbed into the bones will prepare them for smoking. Cold-smoke until the bones are straw-coloured and then either hot-smoke or remove from the smoker and grill as required. If a surfeit of spare ribs is on hand they can all be smoked and then deep-frozen.

Pork can be smoked without being turned into either ham or bacon by using either a 60% brine solution overnight, or an 80% brine for two to three hours depending on the thickness of the cut, or by using an equal mixture of salt and sugar rubbed on and left for 30 minutes. Pork chops can be handled in this way as can gigots or rolled shoulders. After salting, the pieces are cold-smoked and then cooked by hot-smoking, or in a conventional oven or grill.

Pork for paté can be prepared by the above method and then incorporated with the other paté ingredients.

Venison

When considering the smoking of venison, it is advisable to find out about the raw material and to ascertain if it is smokable. When dealing with wild animals the quality of flesh does not create a problem because they only eat when food is available, and only then in ratio to the amount of energy required. Only use young animals and avoid stags near the time of rut. The smaller deer such as fallow, sika, montjack and roe give lighter haunches and other cuts which may be more within the budget of your customers. Farmed animals will display traces of what they have eaten and the quality will therefore be best when a natural diet is used; this should contain not only grass but also materials through which they can browse and pick out bits and pieces they fancy. A spinney with small nut-bearing trees is ideal. Farmed animals should not be fed near the time of killing so that food is no longer present in the gut. Killing with minimum stress is advised because too much adrenalin will be released into the system thus affecting the flesh texture.

There are differences of opinion as to whether a beast should be skinned while still warm but to ensure good colour, I believe that skinning should be done immediately. If, however, it is proposed to use a chill-room for hanging, it may be necessary to use a humidifier to prevent drying the meat too much. Some people prefer to keep the skin on and hang the animal in the chill. Ideally, if you have a game larder which can be kept at around 3°C/38°F then optimum conditions should be present for this process.

When smoking venison there are two options. You can cure and produce a prime quality Parma ham type product or brine and smoke the meat to provide either a cooked meat or a smoked meat ready for cooking. Either are acceptable, and a mix of both types should be tried and a decision made only after experiment. The small off-cuts obviously should be used for pies, stews or paté, sausages or other made-up dishes. In every case great care to keep everything perfectly clean at all stages is essential. Do not use the same cutting board for gutting and dismembering.

To produce your cured smoked venison a tub, large enough to keep the cut covered, is required and a cold room or chill where the container can be kept at around 3°C/38°F. When butchering the carcass try to cut across the muscle as little as possible otherwise the juices will leach out. Try to cut along the natural divisions between the muscle groups. A rough guide to the length of time venison should be left in the cure would be one to one and a half days per 500g/1 lb if on the bone; one day if off the bone. When curing with the bone in, it is an ideal time to use a brine pump. If no brine pump is available take a sharp knife and cut along the bone, easing the flesh away. Pack a little salt into this gap. The amount of salt required should only be a large teaspoonful. The salt will prevent any spoilage which may occur due to the marrow going off. Alternatively, you can bone out without cutting into the flesh by loosening the bone and removing it whole. Use a very sharp, pointed knife and apply as little pressure on the flesh as possible.

The recommended basic cure is as detailed below but the addition of wine, spice or herbs will enable you to produce an exclusive product but remember to take notes to enable you to reproduce your most favourite recipe.

Basic venison cure

40% brine solution
55g/2 oz brown sugar
1 tsp saltpetre

On completion of brining time remove the meat and allow it to drip-dry. Rub very lightly with good quality vegetable oil and cold-smoke for 24–48 hours, depending on preference for strong or mild smoke flavour. Wrap in foil and store in a chill for 24 hours. To mature, wrap the meat in butter muslin and hang in a cool, airy place out of direct sunlight for a week. If storing in a chill avoid over-drying. The longer the time the venison is left before eating, the better the flavour and texture.

Alternatively, a 60% brine solution can be used with the venison being kept in it for 24 hours. Remove and drip-dry. Rub lightly with oil and smoke for 36 hours. Cook either in the smoker or in a ham boiler which will give less weight loss. Cold-smoked joints can be sold for further cooking by the customer.

Goat

Instructions for meat, lamb, and mutton or venison are applicable to goat. Use animals around 12-14 months old and hang for a minimum of 10 days.

Rabbit and hare

Many people may be surprised to find these animals under the one heading but from the point of view of smoking, they have more similarities than differences. The preparation of rabbit is straightforward, the animal being gutted and then well-washed. Many gamekeepers recommend hanging hares for about a week but, in my opinion, this is too long if the hare has to be smoked. Choose plump fresh animals which can either be jointed or left whole for smoking. Both animals are very lean, and, after brining they should be brushed with vegetable oil before cold-smoking. This treatment can be repeated if the flesh appears to be drying out too much. Use a 50% brine solution and leave the rabbit or hare immersed for 12 hours. The following recipe is sufficient for one or two animals and gives a pleasant flavour:

 1 litre/2 pints of water
 170g/6 oz salt
 30g/1 oz demerara sugar
 5 juniper berries
 3 bay leaves
 Pinch of rosemary
 Pinch of tarragon
 Pinch of marjoram
 1 tsp saltpetre

Wash the rabbit or hare thoroughly in cold, salted water to remove all traces of blood before putting it into the brine. Drain, dry and oil the surface (the oil may be mixed with a bouquet of the above herbs). Cold-smoke for 12 hours and raise the temperature to hot-smoke or remove and cook by normal means. If using the smoker for cooking set thermostat at 100°C/220°F and use the 'low heat range' of your smoker (if it has the dual range facility) to give uniform heating without drying out. Alternatively cooking by means of a pressure cooker, ham boiler or roasting the meat in the oven with apples will keep the flesh moist.

Tongue

Fresh tongue of either ox or sheep should be used and a cure prepared in a container. With a fine skewer prick the tongue all over to ensure good brine penetration. Ox tongue is cured for two weeks in a weak brine solution of only 20% to avoid the end-product being too salty. Sheep tongue should be similarly treated for one week. The usual temperature of 3°C/38°F should be maintained throughout the brining time. Add sugar and saltpetre to the brine and pickling spice if a spicy tongue is required in the usual amounts.

Cold-smoke for one to two days keeping the outside moist with vegetable oil. Tongues are usually boiled but may be skinned and baked in the oven, cooking until tender. If you are boiling the tongue, skin it while warm and press it into a round shape in a bowl covered with a plate or use a commercial mould.

Poultry

Poultry is not widely smoked but provided the raw material is suitable, smoke most satisfactorily. A lot has been said and written about salmonella but if basic hygiene is observed and a specific area designated for poultry preparation, the chances of contamination are slight. Salt is used and poultry is always hot-smoked or cooked in another manner to ensure a safe cooking temperature within the bird of 115°C/240°F. Smoked poultry does not tend to dry out in the same way as unsmoked and therefore is suitable as a quick snack or picnic meal. The following recipe, although a little pedantic and suitable for one bird only, is a good training exercise. Once proficiency is gained you will know what you are doing.

    1.5 kg/3 lb chicken
    1.5 litres/3 pints water
    250g/8 oz salt
    140g/4½ oz soft brown sugar
    ½ tsp saltpetre
    Optional: herbs or spice

To prepare chicken for brining, inspect the inside of the bird and ensure that it is clean. Mechanical gutting tends to leave some organs in, particularly the gall bladder and kidney. As these both contain acid they can impart an unpleasant flavour to the flesh if not removed before brining. The salt ruptures the membrane of the organs and releases the acid into the flesh. Cut off the feet, and if you have the patience, pull out the tendons. This is not possible when processing a large number of birds, but does assist the brine intake. Prick the bird all over with a fine needle or skewer to assist brine penetration.

Put the water into a pan and bring to the boil. Add the salt and other ingredients and allow the salt to dissolve. Remove from the heat and allow to cool. Put the chicken in and keep it in the refrigerator or chill at 3°C/38°F for 24 hours. Remove from the brine and drip-dry. If the chickens are to be hung in the smoker, tie them with string under the wings and around the legs to support the weight when placed on the hook. The birds can also be placed on shelves but make sure that they do not touch each other or a pale white patch will be apparent on the finished chicken. Cold-smoke for 12 hours. The hot-smoking gives a darker, more appetising colour to the outside but because the weight loss may reach 20% by so doing, birds are often cooked in a ham boiler or other cooker. To hot-smoke raise the temperature slowly and maintain it at 170°C/350°F until cooked.

When time does not allow for overnight brining use an 80% brine solution with some sugar added and brine the bird for two hours for a 1.5 kg/3 lb 5 oz chicken. Legs, thighs and breast portions should be treated in the same way, but for only half an hour. It is possible to use a 40% brine for the portions as well and leave them overnight. For safety add some saltpetre and sugar in the usual proportion (1 tsp saltpetre and 30g/1 oz sugar per 4.5 litres/1 gallon). Portions will require about eight hours cold-smoking. If selling smoked chickens for further cooking by the customer, a warning label that the birds must be cooked should be attached.

Guinea Fowl

Guinea fowl should be treated as chicken although the flesh tends to be drier. If using a 40% brine allow 12 hours and cold-smoke overnight. Raise the temperature as for chicken and maintain the heat until the flesh is tender. Alternative methods given for any poultry or game bird can be used.

Duck

The best weight for a duck to be smoked is not less than 2 kg/4 lb 8 oz including giblets. A duck of this size has a good flesh to fat ratio. If the breast only is being smoked it is more economic to obtain supplies of a heavy breasted type of bird such as the French variety because they are chunky and give an ample finished product which can be easily sliced.

Select birds fed naturally and allow time for 24 hours hanging once rigor is over. Prepare duck in the same way as chicken, ensuring no organs are left inside. Prick all over with a skewer, especially around the breast. Ducks have a very high fat content and absorb the salt mostly through the body cavity, because the skin is waterproof. It is therefore necessary to make sure the cavity is full of brine and fully immersed. As with all other poultry, treatment depends on the end-product required. Duck may be cold-smoked whole for presenting as roast, smoked duck served hot, or as smoked, cooked duck breast as a starter or as cured duck breast, also as a starter or as an ingredient in canapes.

For whole duck use either an 80% brine solution for two to three hours or 40% brine solution overnight or for at least 12 hours. The weaker brine solution ensures a more even distribution of the brine without making the end-product too salty. Both kinds of brine can have some sugar added (1 tbsp per 4.5litres/1 gallon) to prevent hardening of the flesh.

After brining allow the duck to drip-dry making sure no pockets of brine are left in the cavity. Cold-smoke for 18–24 hours for a strong flavour or for 12–18 hours for a milder smoke. Because of the high fat content it is easier to cook the smoked duck in a bag in a ham boiler to prevent the smoker becoming too greasy, but a short hot-smoke will ensure an attractive colour to the skin without dirtying the machine. The ham boiler ensures less weight loss as well, which is a necessary consideration.

Whole duck can be dry-salted by weighing the duck and making a mixture of half that weight of equal parts of sugar and salt. Any kind of sugar may be used, according to preference. Put the salt and sugar into a bowl and add two teaspoonsful of ground ginger, one tablespoon of soy sauce and two tablespoons of sherry. Mix this together. Rub the inside of the duck with garlic and generously fill the duck cavity with some of the mixture, using the remainder to pat onto the surface of the duck. Leave for one hour and turn the bird over putting the salt/sugar mixture on the underside of the duck. Leave for a further 45 minutes. Wipe off any excess and shake loose the mixture from the inside once you have rubbed in as much as possible. Leave for two hours, then cold-smoke as above. Finish off in the smoker, or by any other cooking method.

Duck breast is a succulent starter which can be cooked or, if cured, served without cooking. The former is more usual but the cured type is an excellently flavoured product with a delicate texture.

Use either an 80% brine solution for two hours or a 40% brine overnight. Additions such as sugar, red wine, spices or herbs should be tried out as they give interesting results. Dry salting can also be used with the proportion being one third of the weight of the bird in salt. Allow to drip-dry or wipe off excess dry salt and cold-smoke for 12–14 hours. Cook in an oven at 144°C/325°F or hot-smoke for the minimum time to make the flesh tender or else the breast meat will be dry.

To cure duck breast use any of the methods suggested above but add one teaspoonful of potassium nitrate per gallon of water as a curing agent. Allow two days in the cure and if the duck is too salty at the end of that time, soak for one hour in cold water. Allow 24 hours settling time before cold-smoking. Smoke for 24 hours and allow a further 24 hours settling time before cold-smoking again. Smoke for a further 24 hours and allow another 24 hours resting time. If the smoke flavour is not strong enough, allow a further 12 hours in smoke. Once smoked, the duck breast has to hang in a temperature of 3°C/38°F for at least two weeks to mature. It is then sliced very thinly and served. Wild duck can be used for any of the above recipes.

Goose

Instructions for duck can be applied to goose successfully. Being much heavier, however, the brining time should be increased to 24 hours for a 40% brine solution and if you are curing goose, allow four days.

Turkey

Today turkey can range from small to very large heavy birds of 12 kg/25lbs and upwards. The weight differential is too great to treat the birds together. Up to 5 kg/11lbs in weight, the instructions for chicken can be followed safely, as they can for joints.

To smoke large turkeys whole it is preferable to use a combination of brine, pump, and solution, otherwise the length of time required for uniform uptake of salt is too long and the bird would become too salty. Unless required for presentation purposes many problems are solved by processing the crown and legs separately.

There are two approaches possible, 80% brine solution for four to six hours depending on size or allow a two-day duration in a 40% solution. When using an 80% brine, inject with a 40% solution three times on either side of the breast and twice on each thigh. Allow at least 12 hours after removal from the 80% solution for the brine to even out. When a 40% brine is being used it is still recommended that the bird is injected with some of the brine. The time allocated does not, however, require to be increased because two days is adequate for the evening-out process to take place. To prevent hardening of the flesh, one teaspoon of saltpetre and one tablespoonful of sugar to 4.5 litres/1 gallon should be added. After brining, allow adequate time to air-dry before cold-smoking because water can collect in the body cavity and be transferred to the smoker, thus increasing the required smoking time. Large whole birds can take as long as 48 hours for the smoke to penetrate throughout the thickness of the flesh and that is why so much care must be given to the brining. Medium birds will only need between 24–36 hours while thighs and crown will be ready in 24 hours. The usual choice of hot-smoking or using other methods of cooking has to be made, but it is always a good idea to give at least 30 minutes hot-smoking as this improves the colour of the finished product.

It is possible to dry-salt crowns and legs but not whole birds above 5kg/11 lb. Take the weight of flesh to be salted and make a mixture of half that weight of equal parts of sugar and salt. Add two teaspoons of saltpetre and mix very thoroughly into the salt/sugar mixture. Line the bottom of a container, place the turkey portions on the bottom and rub the mixture thoroughly on all the meat surfaces before covering with the remainder of the mixture. Leave for 12 hours. Wash off in cold water for 30 minutes and use the times given for cold-smoking as in the previous paragraph. To cook turkey in a conventional oven, use the following guidlines:

Weight range	Temp	Minutes per 500g/1 lb
Up to 4.5kg/10 lb	150°C/300°F	25
4.5-7.25kg/10-16 lb	150°C/300°F	20
7.25kg/16 lb plus	150°C/300°F	18

An alternative method of dry-brining turkey, which is very useful if there is not enough time to do the conventional brine, is to heat half the turkey's weight in salt. Scald the turkey in a pan of boiling water for one to two minutes only. Shake off any excess water and place the bird on a flat surface. Taking care not to tear the skin, pat the warm salt into the flesh and rub as much as the bird will take into the body cavity. Continue until the bird cools. Make a 40% brine solution and inject this once on each side of the breast and twice in each thigh. Leave for two hours. Pat dry and cold-smoke following the recommended times given previously.

Partridge and pheasant

There was a time when to even think of smoking partridge or pheasant was sacrilege, but today, with many more shoots being set up, and the consequent increase in the numbers of birds available each season, smoking is a worthwhile method to add value as well as flavour. The early birds of the season obviously have their market but the older birds, as the season goes on, will be ideal for smoking. If the birds are wild, there is no problem but sometimes intensively reared birds, if not allowed long enough to naturalise in the wild, will still be carrying traces of substances which were included in their feed and this will restrict the uptake of salt. It is advisable to smoke one bird on a trial basis before undertaking batch processing. Birds *must* always be through rigor and they should be carefully inspected for internal organs and lead shot.

The brining of the birds can be a straightforward salt and water or include port wine, red wine, ale, cider or any combination. The basic brine given is for 40% overnight or 75% for one hour. Sugar can also be added in the usual manner.

Any of the suggested brines for poultry can be used including the dry brines, and experiments with spices or herbs give good end-products. After brining and drying birds require cold-smoking overnight or for 12 hours. Lightly rub with vegetable oil when hot-smoking and raise the temperature slowly to to avoid hardening of the outside. Continue hot-smoking until an internal temperature of 71°C/160°F is reached. Alternatively, normal cooking methods can be used or the produce sold as smoked for the customer to cook further. Again, any labelling should state this.

Sometimes birds are badly shot and consequently damaged. The part of the bird that remains useable can be brined and smoked and then incorporated into dishes or made into paté, pies or sausages.

Quail

The quail is a very compact bird but despite this they take a long time to brine and smoke. Care has to be taken not to dry out the flesh as the texture is one of its strong points. Use a 40% brine overnight or 80% for one hour. Rinse and air-dry making sure all moisture is out of the cavity. Cold-smoke for 10–12 hours. Brush with vegetable oil and hot-smoke until a temperature of 71°C/160°F is reached within the bird. The birds can also be finished off by cooking in a pressure cooker or ham boiler and this will prevent high weight loss.

Pigeon breast

Pigeon smoke well but the low amount of flesh on any other part of the bird except the breast is not worth the effort. Pluck the breast and remove the fillet from each side of the bone without removing the skin. Brine in 40% overnight or in 80% for 30 minutes. Cold-smoke for eight to ten hours, lightly brushed with oil. Hot-smoke or cook normally in an oven wrapped in foil.

Wild duck, geese or other birds can all be smoked successfully and recipes for the domestic equivalent will give good results. Those wild birds are usually less fatty than reared birds, and will therefore require slightly less time in brine.

Sausages

Sausages can be made from any raw material such as fish, meat, pork, mutton or poultry. To obtain a good texture a certain proportion of fat is required, and this is usually pork fat. Emulsified protein such as that made from soya bean can be used to stick the ingredients together, but this does not give such a good flavour. The skin of a sausage must allow smoke to penetrate and should either be made of natural casing or manufactured to allow smoke penetration. Natural casings made from treated animal gut are available from butchers' suppliers and come in various sizes to suit small, medium and large sausages. Plastic skins are not suitable. The diversity of sausage types are too numerous to quote here, but the basic method is similar in all cases. It is recommended that you investigate books on sausage-making because this is an ideal way to use up good, scrap, raw material.

Sausage can be described as being ground meat of any variety, seasoned (usually highly seasoned), with herbs, spices and salts. If you are not an experienced 'smoker' sausages are a relatively cheap way to experiment. By curing and smoking a combination of meats to make the sausages, practice is gained along with confidence in the use of the smoker so that you understand how long a temperature change in the smoke can be expected to take. A normal commercial quantity is 45 kg/100 lb of meat, and the amount of salt and spices required would work out as given below:

1 kg/2 lb 3 oz salt
50-100g/2-4 oz black pepper
50-100/2-4 oz sage or other
15-28g/½-1 oz red pepper
15-28g/½-1 oz ground cloves
or
28g/1 oz ground nutmeg

A proportion of two thirds lean meat to one third fat is a good one because it provides a finished sausage that will not lose too much bulk in cooking. With less fat a sausage tends to be hard and dry. Beginner's quantities would be as below:

2 kg/4 lb 6oz meat/fat mixture (one third fat to two thirds meat)
5 level tsp salt
4 level tsp mixed herbs
2 level tsp freshly ground ground black pepper
1 level tsp cloves or nutmeg
1 tsp saltpetre

Trim all the meat from the bones and then trim out all the gristle and blood clots. If the meat is kept cold, trimming and mincing is easier. Cut meat into strips and then into cubes. Cut both the fat and the lean meat in this way, but keep them apart. Weigh the fat and lean meat and mix together in the correct proportion. Mix this with the salt and seasonings and mince this through a coarse plate. Allow to cool and re-mince if desired, using a finer mincing plate.

Stuff the casings and allow it to cure for 24 hours in a cold room or fridge. Cold-smoke for 12 hours or until the sausages are a dark golden colour. Cook as normal.

Slicing sausage

1 kg/2 lb 3oz lean pork (neck or shoulder)
500g/1 lb hard back fat
3 cloves garlic, crushed
2 heaped tbsp salt
1 tsp saltpetre
1 tsp allspice
1½ tsp ground black pepper

Mince the meat once, mix in all the seasonings and the saltpetre. Fill some large casing, making one or two large sausages as you wish. Remember that they will shrink by about one third.

Hang in a steady well-aired temperature of 3°C/38°F for four days. A chill, at the correct temperature, is ideal because it also excludes sunlight. If the temperature in the chill is too low, curing will not take place. Cold-smoke until a deep yellowish-brown colour develops. This will take between 24–36 hours. Hang in a cool airy place for at least a month. When they start to shrink, squeeze them down from each end so that the contents are well-compacted and easy to slice. This can then be eaten thinly-sliced.

Game sausage

The less tender portions of venison can be mixed with lean and fat pork on a 50/50 basis. Instructions and other ingredients are as the first sausage recipe.

Bologna sausage

> 3 kg/6 lb 10oz beef
> 2 kg/4 lb 6oz pork
> 1 litre/2 pints cold water
> 125g/4½ oz salt
> 1 tsp saltpetre
> 2 tsp black pepper
> 1 tsp coriander
> 1 tsp mace onions or garlic if desired

Chill the beef trimmings and mince with salt, using the coarse plate. Add all the other ingredients and cure for 12 hours. Mince the pork with the fine plate, mix in with the minced beef and spices and mince again, using the fine plate. Cure for a further 36 hours. Add the water and mix everything vigorously until the whole becomes sticky. A bowl chopper is useful for this mixing. Stuff the casings tightly and hang for a further 12 hours. Cold-smoke for two hours and then raise the smoker temperature to 42°C/110°F for a further hour. Prepare a pan or boiler with water brought to simmering point. Take the sausages from the smoker and put them into the wate. Cook until the pressure of thumb and forefinger on the casing produces a squeak when released. This will take about 15 minutes. Cool the sausages quickly by transferring to cold water. Hang in a cool place and eat as soon as possible for the best flavour.

Fish sausage

Although not so well-known, fish sausage provides an ideal way for recalcitrant children to increase fish intake in an easy manner. The percentage of water in fish is high, but when salted there is a loss in water content with the remainder forming a paste when the fish is finely minced or liquidised. Fat is needed too and some bulk which might be breadcrumbs or rusk. If the fish is oily, such as salmon or herring, the amount of bulk can be reduced.

The table of proportions given is a guide and experiments with your own levels will give you your preferred end-product.

Fish	65%
Fat	15%
Water	10%
Breadcrumbs or rusk	5%
Salt	3%
Herbs or spices	2%

Carefully remove the bones and skin the fish. Mince. Mix with

Smoking flow chart based on a 90kg/198lb capacity smoker

Key ●●●●●●●●●●●●● Drip drying
▬▬▬▬▬ Brining time
■ ■ ■ ■ ■ ■ ■ Smoking time

One box of sawdust will burn approximately 12 hours.
The day is based on a 9 o'clock start with a refill in late afternoon to allow an overnight smoke

Product	Monday	Tuesday	Wednesday	Thursday	Friday	Saturday	Sunday
1. Cheese	■ ■						
2. Chicken	▬▬▬ ● ● ■ ■ ■ ■ ■						
3. Chicken pieces (legs etc)	▬▬▬ ●▬ ■ ■						
4. Trout whole			■▬● ■ ■				
5. Trout fillet			■▬● ■ ■				
6. White fish			▬▬ ● ● ■ ■				
7. Shellfish			▬▬ ● ● ■ ■				
8. Salmon	● ● ● ▬ ■ ■ ■						▬▬
9. Turkey saddles			▬▬▬▬▬ ● ● ■ ■ ■ ■ ■				
10. Beef	▬▬▬▬▬▬▬▬▬▬▬▬▬▬▬▬ ● ● ● ■ ■ ■ ■ ■ ■						
11. Pork	▬▬▬▬▬▬▬▬▬▬▬▬▬▬▬▬ ● ● ● ■ ■ ■ ■ ■ ■						
12. Lamb	▬▬▬▬▬▬▬▬▬▬ ● ● ● ■ ■ ■ ■ ■ ■						
13. Venison	▬▬▬▬▬▬▬▬▬▬ ● ● ● ■ ■ ■ ■ ■ ■						
Week 2							
First nine products as in week one. Alternate beef and lamb with pork and venison or over a three-week period if greater weight of one product is required.							
11. Pork	▬▬▬▬▬▬▬▬▬▬▬▬▬▬▬▬ ● ● ● ▬ ● ● ▬ ■ ■ ■ ■ ■ ■ ■						

the salt and spices or herbs and the water in a liquidiser or bowl chopper until a paste-like substance is obtained. Mince the fat through a coarse plate and mix this into the fish mixture. Fill the skins and cold-smoke for four to five hours. Cook as for normal sausage. Sausage can also be made from smoked fish and the method is the same. Less time is required for smoking because it will only be to give colour to the sausages.

For those of you considering setting up a business, it is essential that the product you are smoking is processed as efficiently as possible. Many foods require short brining times and many require a long time in the smoker, so by planning a production chart for each type of food, a flow of produce can be effected in such a way that the smoker is in almost constant use. The production chart illustrated here shows how this can be achieved.

To recap, the following step-by-step list when smoking will serve as an aide memoire.

1. Select fresh food and make sure rigor mortis is completed.
2. Weigh.
3. Decide type of brine.
4. Calculate time in brine.
5. Prepare and chill the brine to 3°C/38°F.
6. Put into brine tub and ensure the foodstuff is immersed.
7. Overhaul, turn the meat and stir the brine regularly.
8. Remove from the brine, allow to drip-dry to enable the pellicle to form. Freshwater trout are the exception, and can be put into the smoker wet.
9. Light the smoker with one or two fuel boxes as required.

10. Hang or rack the food and smoke at 21-31°C/70-88°F for the required time, refuelling the sawdust trays if needed.

11. If being further processed in some other way before eating, remove from the smoker.

12. If further cooking is required in the oven, raise temperature and smoke for the required time to finish the food off in hot smoke.

13. Remove from the smoker and allow to cool and mature. The colour and flavour will develop over the first 24 hours after smoking.

14. Wrap and freeze, or tie in muslin and hang depending on product.

15. If required for retail purposes, vacuum pack, or box, the goods and label them.

Chapter 7

Setting Up a Business

NOW THAT smoked food is becoming more and more popular, the desire to become involved in its production, marketing and selling may appeal to a great many enthusiasts, particularly those looking for a new business opportunity or a side business.

However, it is a decision which should be taken with a great deal of forethought and deliberation. Certain people are already well-placed to take advantage of developing a food-smoking business. These are:

• Livestock producers and fish farmers.
• Processors of the produce of the above.
• The catering trade.
• Delicatessens.

All of these businesses already have many of the requirements to enable them to produce smoked food in that they have control of their raw materials and have suitable premises and/or customers.

It does not mean, however, that only they can start and run a successful smoked food business, but they will have addressed vital concerns already before embarking on a business venture such as food-smoking. What foods will be smoked? How large will the operation be? Where will the factory be situated? What is the local competition like?

Raw material is the key to any food processing operation and a good supply at reasonable cost all year round is essential. This can dictate the area in which you will operate because you do not want to incur heavy transportation costs for bringing in raw material or sending off finished goods. When dealing with fresh fish, the needs are even more critical, because landing and handling facilities are needed, and you should be as near to these as possible to avoid the fish having to be frozen.

The idea to smoke food will probably have been initiated by the fact that there is a sufficiency of raw material available, coupled to a ready demand for the finished product in your area, or one of which you know, so your choice of products may already be made for you.

Initially, it is not a good idea to embark on smoking too many different products. You should select products requiring similar treatment and length of time in the smoker, or those which can be processed in the same area. For example, salmon and other fish and shellfish can all be prepared and brined without the need for separate preparation areas. However, if dealing with salmon and chicken, each would have to have a separate area to prevent any possibility of cross contamination.

Another reason for grouping products is that trimming scraps or imperfectly shaped produce can be combined together to make a by-product such as paté, which will eliminate unsold weight, give a better return and hence aid cash flow.

If you produce the raw material you can control the quality of the end-product. Feedstuffs today can be the source of many problems. Do not be tempted to cut corners in the desire for profit straight from start-up. Ensure that the feedstuffs are of sound quality and develop the correct husbandry for the animals from the outset. By maintaining these standards, the end-product quality will be assured and profit will develop accordingly.

If your raw material requires to be slaughtered, the correct control of killing will also enable you to minimise the stress factor when taking animals to the abattoir. The length of time for maturing will also be in your hands, and the flavour and texture will only develop with enough hanging time, as has already been explained. In the case of fish, it will also allow you to organise delivery time to suit your operation, so that the smooth running of the operation is not held up while fish enter or go through rigor.

While deciding what products to smoke the size of the operation will be forming at the same time. If the raw material is of an unusual variety, of which only limited supplies are available, it will not be necessary to have a very large capacity smokehouse. On the other hand, if availability and demand are high, it is obviously an advantage to be able to handle large quantities. The physical size of the commodity may be what determines the size of the smoker and, consequently, the premises in which to house it.

Premises are a major consideration, and when considering them the essentials are water, a road for transport to deliver and uplift, sufficient electricity of the correct phase to power all the equipment you may want to use, drainage, and access doors at either end of the structure or walls which are not too difficult to break through when structural alterations are needed. A beautiful location can be a joy to work in, but it can also be a headache to operate if the road is cut off by snow in the winter or it is so remote that you have to build houses to accomodate the workforce.

Visit, and then remain for a few days in the area in which you are are considering building the factory, and get the feel of the location and the local people. There are taboos about some kinds of work, and if you cannot use local labour and have to import it, you are laying the foundation for future unrest among employees. You need the local population with you.

There is an important issue to be dealt with from the outset when considering a factory. In the event that you might export to the European Union you will have to build, or find a new factory, which will conform to all the environmental health requirements laid down by the EU. Alternatively, find an existing factory and bring it up to the required standards. This decision is usually influenced by finance, and the more compact the premises the cheaper it will be to convert them to conform. Bringing premises up to the required standards today is so expensive that a business can start off with a terrible burden so it is best to check with local enterprise companies to see if there are any viable units on offer.

Early contact with the local environmental officer is essential because interpretation of the EU directive 93/43/AAC:4/6/1993 – L175.19.793 varies and the smoking process is not always fully understood. A meeting at the outset can be helpful to both sides. If you are the first tenant when the unit proposed is being built, changes can be made to drainage and wall finishes which save a lot of money.

If you are a farmer, the probability of being able to convert an outbuilding is high and the drainage system can be altered to meet the required environmental regulations. If you can use your own building, the provision of a shop can be incorporated into the plans and provides a useful marketplace for the end-products.

Hotels, restaurants, and other catering enterprises usually have facilities for preparation and cold storage and often require little alteration to those areas to commence food smoking. These businesses also have existing customers and are therefore in an excellent position to provide something new and different for their clientele without much further outlay.

It is unfortunate that at the time when a potential food smoker requires the knowledge to set up, it is difficult to obtain. It is sometimes easier to seek advice from smoked food consultants who can outline what you need and in what sequence. The initial planning of the factory will ensure efficient, hygienic, working conditions when up to full capacity, where the aim is to have a constant flow of work going forward and no backtracking. This pattern not only prevents muddle but is also means the avoidance of the potential spread of

Flow of Operation in Premises

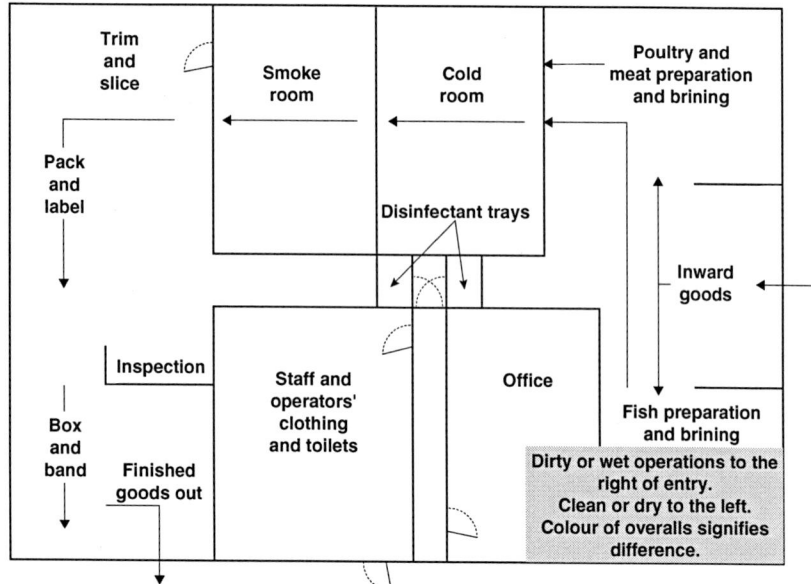

bacteria from one part of the process to another. No matter how small the operation, this must be the aim. Raw material comes in one door, and at the other end the finished product goes out. Cold rooms should have two sections: one for raw material, and the other section for finished goods. Sometimes the premises can be designed so that equipment, such as the smoker and cold room, can be incorporated into the walls, dividing up the premises into sections for each operation of the process. This is only possible where the flow of work focuses around the cold room. Never economise on the size of the cold room you select. Make sure the installation allows you to go in and out with trollies or trays and should, if possible, be without a step up or down.

The diagram illsutrates an ideal layout for a small food-smaoking factory or unit. To be financially efficient the smoker you select should be in operation all the time apart from the hours set aside for cleaning and maintenance. You should also provide space to have a second machine because most factories will have ancillary equipment to support more smoking capacity once the output required justifies the additional capital outlay. Water points should be plentiful and wash hand basins either knee or foot operated. The better the design of the premises, the easier it will be to maintain the required hygiene levels.

When contemplating your choice of factory find out where the discharge goes from each drain. If there is outfall to a river or stream, have the water analysed professionally before you commence operations. This may protect you from any water pollution incident in the future where you may be blamed for pollution which you have not caused.

Premises should always be insect, rodent and bird-proof, so very high old buildings present problems in that you will have to reduce the height of the ceilings to an easy cleaning level. Where possible, select a factory with walls around 2.5m high. Ventilation is a must, and windows will have to have a fine mesh covering to prevent birds or insects flying in. Doors used for loading/unloading should also be provided with screens to keep out pests. An ultra violet insect killer should also be installed and great care taken in the locating of the waste bins which should be under cover; some authorities insist on refrigeration being available to store waste before collection.

A degree of compromise is usually required when selecting premises because if the location is ideal, the building is frequently not, or the other way around. Do not hurry into a decision. The building is the biggest item on your list and involves the major part of your decision-making process. Visit as many factories in the same line of business as you can and talk to the people who work there. Manufacturers of equipment are usually helpful, and there are always improvements being made, so take time to choose what you will install.

The planning stage need not be entirely unproductive because, by starting out with a small food smoker, you can be learning the trade by producing samples and finding customers.

Chapter 8

Finance

Once the decision has been made to set up a smoking business the problem of how to finance the project has to be faced. Innovative ideas can attract a certain amount of practical help, but no one can assist with advice or hard cash until the project has been examined in detail. This entails the costing out of every aspect, from preparing the premises to the supply of overalls for the employees. The drawing up of a business plan can in itself be a costly item, not only in money terms, but also in the time taken to prepare it correctly.

Where can I get help?

If you are planning to take advantage of any Government-backed schemes be prepared for a long delay before you can move ahead, and in the end you may not receive as much as you had hoped for in the way of financial assistance. This delay can finish off a project if it has been based on a firm order and delivery date which the delay then makes it impossible to meet. Also, many people starting up in business may not be aware that grants are usually paid only after a fixed percentage of the proposed grant has already been expended by you. Finance is usually organised as a package so that some money is available as working capital and some as overdraft facility. Each region (or district as it will be in Scotland from April 1996) has an Economic Development Department situated in the head office and this will be able to help you with regard to schemes available for supporting local business enterprise and the grant/funding situation.

If you have already operated a business and are thinking of smoking food as an additional line, talk to your present bankers and ask them what they think. We all make jokes about our bank managers, but if there is a chance for them to lend money on a secured basis, they will be only too glad to make facilities available. Find out all the local schemes for co-operative selling such as the *Food from Britain* scheme or any similar schemes run by the local Tourist Board. These are a good source of assistance for further details not only of the *Food From Britain* scheme, but also in helping you if you have a project which can provide a specific attraction for visitors. If your scheme can be of such interest, the Tourist Board can sometimes assist financially, particularly if a catering facility or reception centre is involved.

Remember, it may not be necessary to go it all alone if your product can complete the range offered by someone else who have already established markets within any such scheme.

Should I raise credit?

Money can always be bought, but credit is expensive, and so it should be determined at the beginning what you intend to make out of the project. Are you wanting to supply local needs on a small domestic scale, or are you wanting to build up a business with a position in the national business community? Once the end-figure is identified, the other costs can be worked out in terms of premises, overheads, output required per day, size and location of the markets and whether your product can support the profit margin you will need to be viable. It always costs a lot more to start a business than you think and the time needed to research potential markets is many months of hard and constant effort when there may be no money coming in to fund the costs.

Preparing a business plan

If you have not been trained to prepare business plans, you should contact an expert who will help you. There are schemes for starting up your own business where this service is part of the package, such as the Enterprise Initiative. Be truthful with yourself about everything, because it is better to recognise and face these difficulties before you start up, than be faced with them when you are actually up and running. If you have a PC, there are bespoke financial software packages which will greatly aid in building up a plan, and these can save a considerable amount of time when fine tuning needs to be done.

How important is the choice of premises?

When making your choice of premises you should allow for the possibility of expansion. It is cheaper to pay a little more rent to begin with than have to move after a few months because there is not enough room. Moves are always very expensive, so the right choice is vital.

Are there any costs which might surprise me?

Another aspect of food production that is often ignored in costings is the labelling of the product. Care must also be taken to ensure that the regulations in force with regard to your product are understood, and that the design and printing of the labels allows the important information to be easily read. Please refer to the illustration on page 108, which carries the legally required information. The cost of this side of the business is very high due to the artwork and design costs involved and the subsequent print bills. It is recommended that you find out the name of any company in your area which is specialised in small print runs of up to 5000 labels and contact them.

You will have to be prepared to give away samples in order to ensure forward orders and establish new customers. Advertising is also expensive and should only be undertaken after a great deal of planning. Be specific and target your potential customers as efficiently as possible. A golden rule in the food production business is to be your own guinea pig and involve your staff, because they will then be more committed to their work. If you happily eat what you make, the chances are a lot of other people will as well.

What sort of return might I expect?

A typical return from a small food smoking business can be expected to come about on the following basis:

Turnover should be approximately three times the cost of raw materials plus direct costs. The gross profit should therefore be twice the cost of materials and direct costs, yielding a net profit of around 50% of the cost of raw materials and direct costs after overheads, salaries and running costs are deducted. In other words, if the cost of your raw materials and direct costs amounts to £10,000 per annum, then turnover should approach £30,000 with a net profit of about £5000. This amounts to a profit margin of 16%.

How can I construct management accounts?

Again, a financial consultant may be able to help as will a chartered accountant. I have always used the layout below to ensure that all aspects of ongoing sales and expenditure are regularly maintained. Financial accounting software for PCs, such as the Sage programs (available from any PC dealer), are ideal in helping set up management accounts and can maintain monthly profit and loss and balance sheet figures.

Sales

Trade
  retail
  wholesale
  exhibitions and fairs
  shop/kiosk
1. Gross turnover   £000

Purchases

  cost of raw materials
  labour
  packaging
  production materials
  hygiene
2. Direct costs   £000
3. Gross profit (1 minus 2)   £000
Overheads
  Training
  Admin salaries
  NI
  PAYE
  Rent
  Rates
  Elec, gas & other
  Tel & fax
  Print & stationery
  Carriage
  Petty cash & postage
  Vehicle running costs
  Repairs & renewals
  Audit & legal
  Insurance
  Leasing
  Bank charges
  Interest
  Loan repayment
  Depreciation
4. Total overheads   £000
5. Net profit (3 minus 4)   £000

Please refer to the Useful Addresses on page 117 for points of contact regarding finance and business start-up.

Chapter 9

Identifying the Market

A smoking operation may be a natural extension to a business already established, and, consequently, the customers will already be identified and a business base ready to build upon. It has often filled me with misgiving, however, when I have been asked to quote for a food smoker of tonnage capacity from a business which has never even smoked a kipper! Enthusiasm is a necessity in any business, but there has to be a realistic side too. Any business needs to know its marketplace, and this will determine whether or not it has a chance of success.

How do you find your market?

The market will most likely be determined by five factors:
1. Your end-product.
2. Your location.
3. Local/national trade potential.
4. Tourist trade potential.
5. Export trade potential.

You will have decided the best end-product to market and your location will be the best one available to you. The local market can be serviced by contacting local hotels, cafes, restaurants, delicatessens, butchers and fishmongers. Do any local shops run 'special' weeks promoting certain foods? Go and talk to some of them anyway, tell them what you would like to do, and offer some kind of special deal. Decide what price you are going to charge, and be consistent with everyone. Never allow the possibility of one customer finding out that another down the street has got a better deal! Work out your pricing carefully, and never be tempted to go below the price you know you need to make the deal viable. Better to do nothing than to supply goods which actually cost you money.

No passing tourist will buy from you if they do not know you exist, or what you make. The best way to remedy this is to have a shop or kiosk where you can show your wares. Local markets are ideal for this because the rent of stalls is usually reasonable, and well within the budget of someone starting out. Find out if there are other open air markets anywhere in your area. Attend local fêtes, agricultural shows and game fairs; remember to give samples as well as offering goods for sale, and always make sure your name and address is clearly displayed and where orders should be sent, faxed or telephoned. A preprinted brochure is essential when attending these types of event. If you are supplying locally, the need to vacuum pack may not be essential, and your customers will then enjoy much better quality, because no matter how well done, vacuum packaging detracts from the flavour of the end-product. Freshly smoked food tastes much better than food kept stored for three weeks.

It is also important to work out your policy on delivery, and whether you will make a charge. Decide on one delivery day for each area in which you supply, and stick to it. Otherwise you may be making a special delivery for a small quantity of salmon.

Advertise a special service such as starters for parties, presents for birthdays and thank you gifts.

In supplying the national market, mail order has become a useful tool for selling, but until people are aware of the quality you produce it is not easy to break into this market. You must also be very sure of your products because it is not the best way to distribute food. Delays in delivery can lead to disappointed customers and replacements which take time and money. Do not ignore it, but leave it for later when the business has grown.

Although supermarket business can be good, do not seek too large a contract before you are sure you can cope at a profitable level. A few bread and butter orders each week to cover the essential outgoings will lay the foundation for growth without overstretching either your financial or production capacity.

Through Government-sponsored trade trips abroad, it may be feasible to attend some of the foreign trade food fairs and in many cases, this can lead to considerable export work. But before you undertake an overseas contract find out all you can about the foreign country involved. In Britain the Department of Trade and Industry have an export payment insurance scheme administered by the Export Credit Guarantee plan which can be found in most major cities. (Your bank will be able to give you the nearest contact address for this or you can contact the local office of the DTI). This body deals with each application on its merit, taking into consideration the commodity and the destination of the consignment, using their knowledge of the area and the probable ability to pay. On the strength of this information they will put forward a proposition giving details of the premium which you can then accept or reject. Have a credit report done on the customer with whom you propose to deal by the DTI who have this facility and for which you pay a small fee, usually about £10. You could also try contacting as many of their other suppliers of whom you have knowledge, to create a profile of the company.

The Export Market Information Centre, run by the DTI is based at Kingsgate House, 66-74 Victoria Street, LONDON SW1E 6SW (tel: 0171-215-5444/5; fax: 0171-215-4231; E-mail: EMIC@ash001.ots.dti.gov.uk) and offers a vast array of services and information on export markets from foreign Yellow Pages, market research reports and trade fair catalogues to street plans of cities.

If you do not speak the language of the country with whom you propose to deal, find an agent who does and is familiar with their methods of trading and any unusual restrictions peculiar to that country. Incorrect documentation can delay payment even on a letter of credit or bank draft, so speak to the overseas branch of your bank and engage their assistance. Do not shun overseas work, but make sure you have everything tied up before you send off the goods, because it is not easy to make redress from a distance and in a foreign language.

Chapter 10

Hygiene and Environmental Matters

When an application for Planning Permission is submitted, Environmental Health Officials are automatically involved. You cannot circumvent this department, so it is far better to ask for their help at the beginning of the project. Once you find a possible factory, sketch out what you think you want in the building and ask an Environmental Health Officer to meet you on site. There is no nationwide policy, at present, on requirements, so by this early meeting you will be made aware of what the local interpretation of the European Guidelines are. (This document exists under reference: L.175.19.7.1993). Through no fault of their own, the officers usually do not have much idea about the smoking process, and are consequently on the defensive, but will be willing to co-operate. They have to justify their existence, but also, they want projects to go ahead to bring employment to the area, so if you make it plain you want it set up according to the guidelines and regulations, but are not willing to drain off cash in the process, you should develop a business understanding and a working relationship.

The law lays down certain provisions with which you must comply, such as a rest room, separate male and female toilets, canteen facilities, a disabled toilet which will take a wheelchair (though this can be provided as an oversized female toilet). Without discrimination, most food processing business involves lifting boxes of raw material, and as such is rarely suitable employment for physically disabled people. It is, of course, possible that some job might be without hazard to a disabled person, and therefore, provision should be made. You must have proper insurance for all aspects of the business including premises, equipment, products, employers' liability, and a good lot more. If you are not already familiar with these requirements talk to an insurance consultant, and then shop around for quotations as charges vary considerably.

If your factory is carefully planned it will eliminate the possibility of trailing dirt from one area into another. The factory should be divided physically into wet/dry or clean/dirty areas with no interaction between the two sections able to take place.

Food hygiene is more than just cleanliness; it includes all the practices involved in:

1. Protecting food from risk of contamination, including harmful bacteria, poisons and foreign bodies.

2. The prevention of any bacteria present multiplying to an extent which would result in the illness of consumers or result in early spoilage of the food.

3. Destroying any harmful bacteria in the food by thorough cooking or processing or issuing potential consumers with instructions which will do so.

Food is a potentially dangerous commodity, and must always be treated with the utmost care. Food poisoning is a serious illness, and the more you know the easier it is to avoid circumstances which may cause an outbreak. There is a group of foods considered high risk foods, and those are listed below:

1. All cooked meat and poultry.
2. Cooked meat products including gravy and stock.
3. Milk, cream, artificial cream, custards and dairy produce.
4. Cooked eggs and products made from eggs, for example, mayonnaise.
5. Shellfish and other seafoods.
6. Cooked rice.

Food poisoning resulting from raw materials can be caused by:

1. Bacteria or their toxins.
2. Viruses.
3. Chemicals such as insecticides and weedkillers.
4. Metals such as lead, copper, and mercury.
5. Poisonous plants such as deadly nightshade and toadstools.

As a food processor you could cause an outbreak of food poisoning due to any of these factors but the most likely of all is bacteria and their toxins.

Microbes are tiny, invisible, living organisms of which there are three kinds which food handlers should know about.

1. Bacteria.
2. Yeasts.
3. Moulds.

Bacteria are commonly known as 'germs'. Under a powerful microscope they are seen to be round or rod-shaped and some have strands which help them to move easily in water. They are found everywhere: in dust, dirt, water, air, soil and on people, animals, food, insects, clothing, etc. Most bacteria are not harmful to people and some are used in the production of food, e.g. yoghurt. Certain bacteria cause breakdown and decay. Others can make poisons which create disease. Usually there must be quite large numbers present to cause this.

Yeasts are very small plants which are found in nature, e.g., on the skins of fruit such as grapes. They produce spores which can live in dust, soil and dry foods, and because they are so light they can be carried about in the air. When yeasts land on moist foods they begin to react with the natural starches and sugars. As they do so, carbon dioxide and alcohol are given off. It is because of this that certain yeasts are specially grown to be used in bread-making and others used in wine-making, brewing and distilling.

Moulds are known to us all! Like yeasts, they are tiny plants. Growing moulds can be easily seen as patches of colour on over-ripe food (such as fruit) — white, green, grey and orange. These are really groups of millions of 'seeds' called spores, which are so light that they too are carried away with the smallest movement of air. The spores can be found in dust, air, soil, paper, on clothing, people, animals, and all kinds of equipment and articles. When the spores land on a suitable place to grow, they make a network of strands which digest the host food. Moulds can feed on a very wide range of things: wood, paper, leather, human foods of all kinds, even living plants and animals. Most moulds cause decay but not disease. Some are used in food production, e.g., 'blue' cheeses. Other types produce antibiotics such as penicillin.

All of the above are frequently implicated in outbreaks of food poisoning, especially with poultry and cooked meats. Unfortunately, contaminated food usually looks, tastes and smells completely normal. One advantage to food smoking is the fact that salt is used to such a great extent, and this does help to prevent bacteria multiplying.

The temperature chart opposite showing the effect on bacteria will clarify what happens if bacteria are present and not dealt with. Only proper cooking to the correct *internal* temperature can enable food to be eaten safely.

When dealing with smoked foods only the raw material from proven uncontaminated sources may be cold-smoked. This applies mostly to fish and shellfish. All freshwater fish including trout must be either hot-smoked after cold-smoking or cooked in some other way. When dealing with poultry, game and meat, unless the product has been cured or pickled it is not tender enough to eat and therefore further cooking must be done. Instructions on the label must always be clear when cold-smoked products are offered for sale, because unless clearly stated that further cooking is required, it can lead to the idea that all smoked foods are ready for immediate consumption.

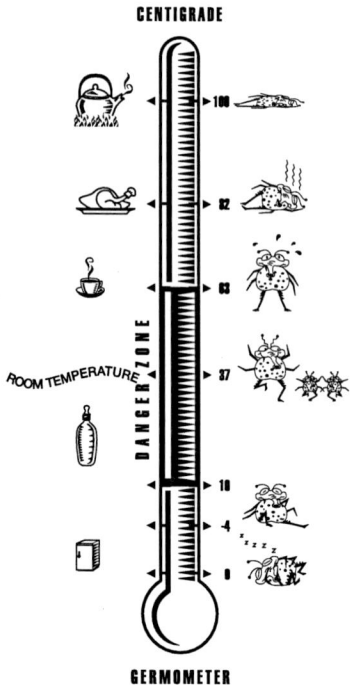

CENTIGRADE

GERMOMETER

In order to deal with the possibility of food poisoning, prevention by the producer and employees is the first and most vital step. Health regulations tend to annoy individuals but when operators understand that a single, careless action by an individual can close an operation down and risk their job as well, a responsible attitude can be fostered. It has already been explained that the correct planning of the factory to ensure no back-tracking or connection between the various parts of the process will lessen the possibility of cross-contamination. It is now law that all food handlers hold the basic food hygiene certificate (No: 7025) and this is the first stage in awareness.

Many people believe that overalls are worn to protect the individuals clothes but is primarily to prevent the spread of bacteria from their own clothes or person. The use of knee or foot operated taps on wash hand basins also helps to maintain hygiene levels. Wash hand basins should be placed near the operators workplace, because they will then not object to using them.

It is a legal requirement to record the daily temperature of all refrigerators and cold rooms and this should be the first job of the day for the person responsible. Records will be maintained in different ways in each factory, but they are vital. Even if all records are computerised, a daybook where facts are entered in situ can then be used to compile more sophisticated data records. Apart from the temperature, all incoming raw material should be given a batch number which will follow it through the whole process and finally appear on the package sent out for sale. This gives a record of all that has happened to this raw material and can, if routine analysis is done, identify any area in which something is wrong. These analyses will give a breakdown of the levels of bacteria, nitrite, nitrate and salt and if any of these are above the agreed levels for your product, the offending batch is identified and can be withdrawn if required. Notes on the appearance of the end-product should also be made in the daybook.

When smoking food the weather should also be recorded because if the smokehouse is not fully dehumidified the actual smoking process can take longer. When it is raining the air entering the smoker is wetter and is less able to draw off the moisture from the food. Antiseptic foot baths, changed frequently, will stop dirt from one area being taken into another. It has been argued that when using trolleys this makes it difficult to wheel them about, but it also ensures that the wheels of those trolleys do not spread contamination about. If the baths are in natural doorways a slight slope can be constructed at either side in concrete and this makes the operation less difficult.

The undernoted list will simplify what is needed and if it is adhered to, very little can go wrong.

1. Chilled food must be kept below 10°C/50°F.

2. Freezer temperature should be below -18°C/0°F.

3. Bacteria are not destroyed by chilling or by freezing.

4. When cooking, food must reach an *internal* temperature in the middle of the food of 70°C/158°F. The item must be cooked long enough for every part of it to reach at least this temperature.

5. If keeping food hot it must be kept above 63°C/145°F.

6. Remove waste from working surfaces into a bin or bowl designated for this purpose.

7. Empty this frequently and wash it.

8. Use the waste disposal unit for soft food waste (boned waste which can be ground down).

9. Have plastic bins with liners to collect waste matter not suitable for the waste disposal unit.

10. Empty the bins regularly and wash if necessary.

11. Have a strong dustbin/skip outside.

12. Keep the outside refuse area clean and tidy.

13. Use tight-fitting lids for dust bins.

14. Keep clear of animals and birds.

15. Wash hands if putting waste in bin.

16. Strictly no smoking other than in designated areas in rest rooms.

Chapter 11

Packaging and Labelling

IT IS very satisfying to look on freshly produced food, and to know it tastes good. This should be enough, but today we insist on going to great expense to envelop good food, and then put a label on to say what is inside the package!

The market into which you will sell your product will dictate the type of packaging, but unless you are transporting food for more than two or three hours, it is possible to use methods other than vacuum packing.

A review of some possible packing methods is given below and the more elaborate, giving longer shelf-life and more protection, come towards the end of the list.

1. Straight from the counter to a plastic bag.

2. Shrink-wrapped, either with or without heat-sealing.

3. Shrink-wrapped in plastic or waxed cardboard trays, or on a backing of similar material.

4. Packaged in a good quality plastic pouch and heat-sealed.

5. Vacuum packed in an impervious plastic pouch, directly or on to a backing of plastic, waxed card or aluminium foil.

6. In a plastic impervious pouch which is evacuated, partially filled with an inert gas (nitrogen) and then heat-sealed.

7. Retort pouch which is made of plastic and aluminium foil as a three-ply impervious material and is claimed to give a shelf-life of two years without refrigeration. Pouches are also made without the aluminium foil, are transparent, but have a shorter shelf-life.

8. Jars (for small items in preserving fluid).

9. Plastic tubs, as in 8 above.

The information that must be given when labelling is extensive and leaves little room for decoration.

You must state:

1. The name of the food.

2. Ingredients.

3. Net quantity/weight.

4. Datemark.

5. Name and address of manufacturer or packer.

6. Place of origin (when necessary).

7. Instructions for use when necessary (i.e., cooking instructions).

8. Any special conditions for storage or further use.

When you are thinking of ways to package your goods, go round all the shops which sell quality food and really look at the packaging. When you examine it closely, it is surprising how many flaws you can find, and, in the end, the simple packages are usually the most effective. The designing of labels can be difficult to fit everything on and the list of E numbers which are allowed cuts down the space available. (See Appendix A.)

Labelling laws are at present in a state of flux and local environmental officers are not in agreement about the interpretation of the directives from the EU. Establish a good relationship with your labelling officer and you can then work out a compromise. He is the man who has to fight for you if someone takes you to task, claiming an infringement of the directive, so he has to feel happy with what your labels state. When agreeing the details on a label with your environmental officer, it is wise to explain exactly what is in the product because you will be surprised at what is and is not permissible. How you describe a process may have to be reworded to comply.

The illustration shows the kind of layout you may choose, but when designing a label a decision as to whether you will require a bar code must be made. The larger food organisations insist on bar codes and this does involve expense as each commodity has to have a separate code and each costs in the region of £70. (Computer software now exists which allows the generation of your own bar codes, but care should be taken that the package is applicable for your product and retail base). If you feel that at some time in the future

you may need these, then leave space in your design to incorporate the feature.

Labels today are printed in a size and format to fit in with an electronic weighing scale/labeller. When selecting this essential piece of equipment you must consider the number of products you may want to label and the levels of pricing involved. *Do not hurry* when deciding on the make of labeller you need. It is perhaps the most critical purchase you will make and being able to amend the information on the label easily will prevent tempers being frayed when you are busy. The manufacturers of weigh/labelling equipment are helpful and your local representative will tell you what the machine can or cannot do, so listen to what he says and make sure it is the machine for you.

Early on when deciding how you will pack and label, it is a very good idea to spend money employing a food analyst. You should ascertain the levels of substances that are allowed and not allowed, the levels of nitrates and nitrites allowed and also the permissible bacterial levels. The raw material should also be analysed before you start processing and this helps determine whether there are any areas in the process where contamination might be picked up.

Once you are happy with your product you have to decide the shelf-life you will give it and only by analysis can you determine how long the product will last before levels are too high to allow your product to be sold. The longer products are stored, the more apparent will be the increase in bacteria, nitrites and nitrates. Samples taken at varying lengths of time from the packing date will give you the answer. It is also a good way to establish if any changes in taste occur over a period of time which might make the product unpalatable. Routine checks on products are recommended as well as staff tasting panels to ensure there is no dropping on standards.

Chapter 12

Dos and Don'ts in Food Smoking:

1. Raising the temperature too quickly will 'case-harden' the exterior of the flesh and prevent proper penetration of smoke.

2. Too high a temperature when smoking fish will make them disintegrate, and if hung by the head they will collapse.

3. Always remove the kidney from eels. The kidney lies about 25-40mm/1–1½ inches below the vent.

4. Ensure proper weight loss takes place for best results.

5. Allow adequate time for drying before smoking as this effects the look of the finished product.

6. In poultry and game birds check that the gall bladder has been removed. If salt reaches the gall bladder it ruptures the membrane, releasing acid into the flesh with a consequent unpleasant flavour and smell. The kidneys and all other organs and traces of blood should be similarly washed out.

7. When smoking eels and trout, especially farm fish, try to ensure that an adequate 'starvation period' has taken place before killing; otherwise, the eels may taste musty or the flesh of the trout be very soft and oily.

8. When smoking previously frozen salmon, if a cooked fish is required, it is helpful to rub the surface of the fish with a little oil before raising the temperature.

9. When smoking lean meat or venison, a little water in the drip tray or in a small pan placed in the oven prevents excess drying.

10. Beef, pork and venison must age a minimum of one week before brining and smoking.

11. Do not use damp or wet sawdust or a mouldy flavour may be imparted to the product and drying will take much longer. Avoid sawdust from newly-felled trees.

12. When hot-smoking, experiment by adding some twigs and leaves of the fruit woods such as apple, pear and cherry to the sawdust. Gorse, heather, hawthorn and rose wood can also be used

along with juniper berries. Although these do not effect internal changes, the aroma of the meat can be changed in a most attractive manner.

13. Using resinous wood shavings and sawdust gives a bitter flavour. The Finns like it but it does not usually appeal to the average British palate.

14. Keep the temperature of long-term brines at around 3°C/38°F.

15. Brine the foodstuff for the specified time.

16. Keep a separate brine tub for each type of food and it can then be in use all the time.

17. If the smoker is full of food make sure that no pieces are touching because this prevents proper circulation of smoke.

18. Do not smoke fish with any other type of food. If unavoidable, keep to one side and avoid drips.

19. If no other suitable hanging space is available the smoker may be used for pre-drying. Always remove the drip tray and dry it before lighting the fuel.

20. Remove scent glands from small game animals to prevent strong animal smells.

Appendix A

E Numbers

E322	Lecithins
E325	Sodium lactate
E326	Potassium lactate
E327	Calcium lactate
E330	Citric acid
E331	Sodium dihydrogen citrate
E331	diSodium citrate
E331	triSodium citrate
E332	Potassium dihydrogen citrate
E332	triPotassium citrate
E333	Calcium citrate
E333	diCalcium citrate
E333	triCalcium citrate
E334	Tartaric acid
E335	Sodium tartrate
E336	Potassium tartrate
E336	Potassium hydrogen tartrate
E337	Potassium sodium tartrate
E338	Orthophosphoric acid
E339(a)	Sodium dihydrogen orthophosphate
E339(b)	diSodium hydrogen orthophosphate
E339(c)	triSodium orthophosphate
E340(a)	Potassium dihydrogen orthophosphate
E340(b)	diPotassium hydrogen orthophosphate
E340(c)	triPotassium orthophosphate
E341(a)	Calcium tetrahydrogen diorthophosphate
E341(b)	Calcium hydrogen orthophosphate
E341(c)	triCalcium diorthophosphate
E400	Alginic acid
E401	Sodium alginate
E402	Potassium alginate

E403	Ammonium alginate
E404	Calcium alginate
E405	Propane-1, 2-diol alginate
E406	Agar
E407	Carrageenan
E410	Locust bean gum
E412	Guar gum
E413	Tragacanth
E414	Acacia or Gum Arabic
E415	Xanthan Gum
E420(i)	Sorbitol
E420(ii)	Sorbitol syrup
E421	Mannitol
E422	Glycerol
E440(a)	Pectin
E440(b)	Pectin, amidated
E450(a)	diSodium dihydrogen diphosphate
E450(a)	tetraSodium diphosphate
E450(a)	tetra Potassium diphosphate
E450(a)	triSodium diphosphate
E450(b)	pentaSodium triphosphate
E450(b)	pentaPotassium triphosphate
E450(c)	Sodium polyphosphates
E450(c)	Potassium polyphosphates
E460(i)	Microcrystalline cellulose
E460(ii)	Powdered cellulose
E461	Methylcellulose
E463	Hydroxypropylcellulose
E464	Hydroxypropylmethylcellulose
E465	Ethylmethylcellulose
E466	Carboxymethylcellulose, sodium salt
E470	Sodium, potassium and calcium salts of fatty acids
E471	Mono- and di-glycerides of fatty acids
E472(a)	Acetic acids esters of mono- and di-glycerides of fatty acids
E472(b)	Lactic acid esters of mono- and di-glycerides of fatty acids
E472(c)	Citric acid esters of mono- and di-glycerides of fatty acids
E472(d)	Tartaric acid esters of mono- and di-glycerides of food fatty acids
E472(e)	Diacetyltartaric acid esters of mono- and di-glycerides of fatty acid
E473	Sucrose esters of fatty acids
E474	Sucroglycerides

E475	Polyglycerol esters of fatty acids
E477	Propane-1, 2-diol esters of fatty acids
E481	Sodium stearoyl-2-lactylate
E482	Calcium stearoyl-2-lactylate
E483	Stearyl tartrate

Appendix B

Useful Addresses and Points of Contact

Small firms Loan Guarantee Scheme

SSLGS
c/o Moorfoot
St Mary's House
SHEFFIELD S1 4PQ

Local Enterprise Companies

These are contactable through your local phone book.

Enterprise schemes

These will fall under the remit of the above who will probably have knowledge of other non-governmental shemes in operation in your area.

Venture capital

This is available from organisations such as 3i (Investors in Industry) and other companies throughout the UK. Local Enterprise Companies can advise, as can your own bank. Thorough and professional advice (such as your chartered accountant can offer) before approaching these organisations is essential and a properly prepared business plan is a prerequisite.

UK Trade organisations

Your local Chamber of Commerce will be able to put you in touch with potential suppliers and customers as well as giving you information regarding overseas Chambers of Commerce members. Joining your local Chamber of Commerce is good value and if you use their services as often as possible, the annual subscription is quite justifiable.

Export intelligence

SCOTTISH INNOVATION
Templeton Business Centre
GLASGOW
G40 1DA
Scottish Innovations does not supply finance, but provides international business opportunities and leads delegations of Scottish companies to major business partnership events throughout Europe.

Local Chambers of Commerce can also help in making contact with potential overseas trading partners/customers etc.

Suppliers & ancillary equipment manufacturers

Check local Yellow Pages under Kitchen Equipment, Catering Equipment.

Smoke oven manufacturers

AFOS
Unit 5, Kingswood Business Park
Kingswood
HULL HU7 3AP
ENGLAND

FESSMANN
Postfach 360
D-71351 WINNENDEN
GERMANY

INNES WALKER
14 Macadam Place
South Newmoor Industrial Estate
IRVINE KA11 4HP
SCOTLAND

MAURER
22 Greenbay Road
WINNETKA, IL 60093
USA

VEMAG
PO Box 1620
D-VERDEN (ALLER)
GERMANY

Index